# Intermittent Water Supply: Challenges, Opportunities and Solutions

*In Focus – Special Book Series*

# Intermittent Water Supply: Challenges, Opportunities and Solutions

Editors

Raziyeh Farmani, Bambos Charalambous, Kara L. Nelson
and Joe Dalton

Published by          **IWA Publishing**
                      **Unit 104-105, Export Building**
                      **1 Clove Crescent**
                      **London E14 2BA, UK**
                      Telephone: +44 (0)20 7654 5500
                      Fax: +44 (0)20 7654 5555
                      Email: publications@iwap.co.uk
                      Web: www.iwapublishing.com

First published 2023
© 2022 IWA Publishing



**Disclaimer**

The information provided and the opinions given in this publication are not necessarily those of IWA and should not be acted upon without independent consideration and professional advice. IWA and the Editors and Authors will not accept responsibility for any loss or damage suffered by any person acting or refraining from acting upon any material contained in this publication.

*British Library Cataloguing in Publication Data*
A CIP catalogue record for this book is available from the British Library

ISBN: 9781789064247

# Contents

doi: 10.2166/aqua.2023.001

# Editorial: Intermittent water supply: challenges, opportunities and solutions

Intermittent water supply (IWS) is a pervasive issue in many countries across South Asia, Latin America, Africa, and the Middle East. Various factors can lead to intermittency, including natural, technical, and financial scarcity, user behaviour, and institutional aspects. Unfortunately, intermittent access to water can negatively impact public health and social equity, and the systems themselves are often considered unsustainable in the long term.

To address this issue, researchers and practitioners have been working to better understand IWS systems and improve their operation, management, and eventual conversion to continuous water supply systems. This special issue presents a collection of high-quality, peer-reviewed technical papers that address the challenges, opportunities, and solutions of IWS systems.

The papers in this special issue cover a range of topics, including:

- Role of social, economic, technical, and institutional factors in the success or failure of IWS systems (Satpathy & Jha 2022),
- Methods and tools for modelling intermittent water supply systems (Sarisen *et al.* 2022; Sinha *et al.* 2023),
- IWS network management (Erickson *et al.* 2022; Ferrante *et al.* 2022) and
- Water quality management (Rukshar *et al.* 2023).

Authors contribute to highlighting the importance of these topics and providing innovative solutions to the challenges facing IWS systems.

Overall, this special issue is an important contribution to the ongoing efforts to address the challenges of IWS systems, and the papers within it provide valuable insights for researchers and practitioners.

**Guest Editors**

**Raziyeh Farmani** IWA
Centre for Water Systems, Department of Engineering, University of Exeter, Exeter, UK (r.farmani@exeter.ac.uk)

**Bambos Charalambous** IWA
Hydrocontrol Ltd, P.O. Box 71044, Limassol, Cyprus (bcharalambous@cytanet.com.cy)

**Kara L. Nelson** IWA
Department of Civil & Environmental Engineering, University of California, Berkeley, California, USA (karanelson@berkeley.edu)

**Joe Dalton** IWA
HydrOptimise, MAZ Business Development, The Lagoon, Amwaj, Bahrain (joe.dalton@gmail.com)

## REFERENCES

Erickson, J. J., Nelson, K. L. & Meyer, D. J. 2022 Does intermittent supply result in hydraulic transients? Mixed evidence from two systems. *AQUA – Water Infrastructure, Ecosystems and Society* **71** (11), 1251–1262.

Ferrante, M., Rogers, D., Mugabi, J. & Casinini, F. 2022 Impact of intermittent supply on water meter accuracy. *AQUA – Water Infrastructure, Ecosystems and Society* **71** (11), 1241–1250.

Rukshar, Vyas, A. D. & Bhatnagar, N. 2023 Assessing the deteriorating water quality in wards of Jaipur city through GIS interpolation. *AQUA – Water Infrastructure, Ecosystems and Society* **72** (4), 425–437.

Sarisen, D., Koukoravas, V., Farmani, R., Kapelan, Z. & Memon, F. A. 2022 Review of hydraulic modelling approaches for intermittent water supply systems. *AQUA – Water Infrastructure, Ecosystems and Society* **71** (12), 1291–1310.

Satpathy, S. & Jha, R. 2022 Intermittent water supply in Indian cities: considering the intermittency beyond demand and supply. *AQUA – Water Infrastructure, Ecosystems and Society* **71** (12), 1395–1407.

Sinha, A. K., Ghorpade, A., Damani, O. & Kalbar, P. P. 2023 Hydraulic modeling approach for evaluating the performance of flow-starved water transmission networks. *AQUA – Water Infrastructure, Ecosystems and Society* **72** (1), 1–18.

doi: 10.2166/aqua.2022.206

# Does intermittent supply result in hydraulic transients? Mixed evidence from two systems

John J. Erickson [a,*], Kara L. Nelson [a] and David D. J. Meyer [b]

[a] Department of Civil and Environmental Engineering, University of California, Berkeley, CA 94720, USA
[b] Department of Civil and Mineral Engineering, University of Toronto, 55 St. George St., Toronto, ON M5S 0C9, Canada
*Corresponding author. E-mail: john.j.erickson@gmail.com

JJE, 0000-0001-6776-5444; KLN, 0000-0001-8899-2662; DDJM, 0000-0003-0979-118X

## ABSTRACT

Pressure transients can cause severe damage in continuous water supply pipe networks, but little is known about pressure transients in intermittent networks. Published examples of high-frequency pressure monitoring in intermittent networks are lacking. Intermittent supply can be caused by poor network condition and is associated with delivering less water, less frequently, and with poorer quality than continuous supply. Given the frequency with which intermittent systems drain, fill, and change supply regimes, pressure transients have been hypothesized to be common and to be one mechanism by which intermittent supply further degrades network condition. We present supply start-up data from two very different intermittent systems: a low-pressure, intermittent network in Delhi, India, and a higher-pressure intermittent network in Arraiján, Panama. Across monitoring locations at both sites, we did not detect substantial pressure transients due to pipe filling. In Arraiján, pump start-ups, pump shutdowns, and pipe bursts were associated with potentially problematic transients. We conclude that pipe filling in intermittent supply does not always result in concerning pressure transients. The largest risks to pipe conditions we observed were due to pumping changes in close succession; hence, we recommend that utilities operating intermittent (and continuous) systems leave adequate dissipation time between changes in pump operation.

**Key words**: hydraulic transients, intermittent water supply, pipe damage, pipe filling, pressure monitoring

## HIGHLIGHTS

- High-frequency pressure monitoring was conducted in two intermittent drinking water distribution networks.
- We found no evidence of substantial pressure transients due to pipe filling.
- In Arraiján, pump starts and stops were associated with transients.
- Pump starts and stops in rapid succession posed a greater risk to pipes than pipe filling did in the observed study zones and should be avoided in intermittent systems.

## GRAPHICAL ABSTRACT

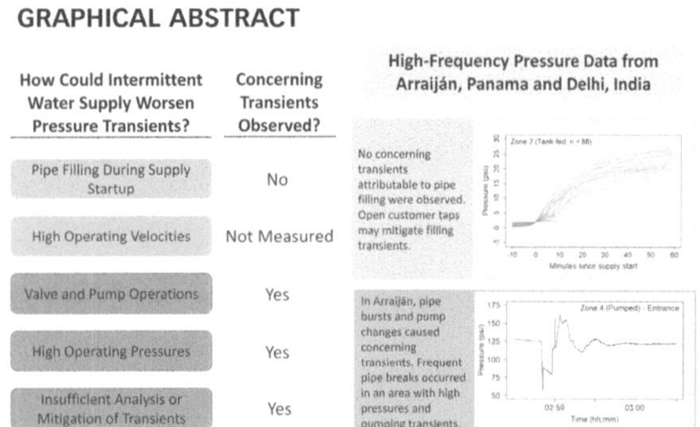

## 1. INTRODUCTION

It is estimated that more than 1 billion people in low- and middle-income countries worldwide receive water from piped drinking water systems that are available to users for an average of less than 23 h/day (Bivins *et al.* 2017). Intermittent supply inconveniences users (Lee & Schwab 2005; Cook *et al.* 2016; Erickson *et al.* 2020), threatens water quality (Coelho *et al.* 2003; Lee & Schwab 2005; Klingel 2012; Kumpel & Nelson 2016), and leaves some users with insufficient volumes of water (Andey & Kelkar 2007; Kumpel *et al.* 2017; Taylor *et al.* 2019).

Intermittent water supply (IWS) can be caused by insufficient water resources, inadequate infrastructure, unplanned expansion of the distribution network, excessive water losses, ineffective operational strategies, or a combination of these factors (Yepes *et al.* 2001; Vairavamoorthy *et al.* 2007; Klingel 2012; Galaitsi *et al.* 2016; Kumpel & Nelson 2016). Many have argued that the hydraulics of IWS worsen an intermittent network's condition, causing a potentially vicious cycle (Yepes *et al.* 2001; Lee & Schwab 2005; Charalambous 2011; Christodoulou & Agathokleous 2012; Klingel 2012; Galaitsi *et al.* 2016) in which IWS causes pipe breaks and leaks, resulting in increased water losses and more severely intermittent supply. But due to the complexity of IWS, little rigorous evidence is available to evaluate if, where, or which specific IWS conditions cause pipe degradation.

Hydraulic transients, commonly known as a 'water hammer' (Boulos *et al.* 2005), are purported to be one component of the vicious cycle of intermittent supply (Lee & Schwab 2005; Klingel 2012). But to date, the influence of intermittent operations on the severity and/or frequency of hydraulic transients remains unknown. Hydraulic transients can result from any sudden disturbance in pressure or flow conditions, such as during pipe filling, pump start-up or shutdown, valve opening and closing, rapid changes in demand (such as opening or closing of large customer connections), entrapped air, or pipe breaks (Boulos *et al.* 2005). As such, the severity and/or frequency of hydraulic transients could be worsened by IWS because:

1. Many intermittent networks regularly fill. Filling pipes can trap air pockets, which can induce transients when they collapse rapidly under increased pressure (Izquierdo *et al.* 1999); when the last of the air is expelled through an orifice (such as an open tap) and water begins to exit the orifice, suddenly increasing the hydraulic resistance and reducing the filling velocity (Li & Zhu 2018); and/or when air release valves close suddenly after expelling the air (Batish 2003; Lingireddy *et al.* 2004). Weston *et al.* (2022) measured significant hydraulic transients during the filling of a partially drained pipe in the laboratory designed to mimic many aspects of intermittent supply.
2. Many intermittent networks regularly change supply regimes. Changing supply regimes, at least in continuous water supplies, can cause transients due to pump shutdowns (Geldreich 1996) and valve operations (Batish 2003; De Marchis *et al.* 2010; Kumpel & Nelson 2014) – both of which are common under IWS.
3. Intermittent networks supply demand over shorter time intervals than continuous networks, which may result in higher flow velocities because most IWS networks were not built to operate intermittently (Ghorpade *et al.* 2021). All else being equal, higher flow velocities result in more severe transients (Karney & McInnis 1990).
4. Some intermittent networks have high pressures, especially near sources. Unplanned network expansion may necessitate the expansion of pumping capacity to force greater flows through undersized pipes, leading to excessive pressure near the pump station. Transient pressure increases superimpose on steady-state pressures, compounding the risk they pose to infrastructure.
5. Many intermittent systems are operated without accurate knowledge of the system (Klingel 2012). Uncertainty in operating and loading conditions prevents the effective analysis (and thereby mitigation) of transients (Karney & McInnis 1990).

Yet the severity and/or frequency of hydraulic transients could be mitigated by IWS because:

1. Intermittent systems tend to have high leakage rates when they are pressurized (Yepes *et al.* 2001; Klingel 2012; Galaitsi *et al.* 2016). Leaks act as pressure-dependent demand, helping to dissipate the energy from (and therefore severity of) hydraulic transients (Karney & Filion 2003). Furthermore, in some intermittent systems, users leave their taps open before and during supply, creating additional pressure-dependent demand (Vairavamoorthy *et al.* 2007) that can dissipate hydraulic transients.
2. Intermittent systems often have large pressure losses in the network, which reduce the average network pressure (Lee & Schwab 2005; Vairavamoorthy *et al.* 2007; Kumpel & Nelson 2016). As transient pressures are superimposed on network pressures, lower network pressures reduce the risk of over-pressurizing pipes (but increase the risk of contaminant intrusion (Taylor *et al.* 2018)).

Few studies have measured pressure in intermittent systems at a high enough frequency to characterize hydraulic transients. In one previous study, pressure was monitored during 16 supply cycles (at 4 Hz) in intermittent portions of a distribution network in Hubli-Dharwad, India, where system operators opened and closed valves during the supply cycles (Kumpel & Nelson 2014). One event was classified as a 'transient,' but as the observed pressures were between 0 and 15 psi, such an event would be very unlikely to damage pipes. Several other studies reporting pressure in IWS sampled too infrequently to detect transients, measuring every 5–30 min (Andey & Kelkar 2007; De Marchis *et al.* 2010; Al-Washali *et al.* 2018; Campisano *et al.* 2018; Meyer *et al.* 2021; Sánchez-Navarro *et al.* 2021).

Despite the prevalence of IWS and concerns about its effects on pipe infrastructure, there are critical knowledge gaps regarding the severity and mechanisms of pipe damage that occurs under IWS. Given the heterogeneity of intermittent networks, the goal of this study was to evaluate the prevalence of extreme pressures and pressure transients in five contexts: (i) a normally high-pressure continuous supply that experienced 11 unplanned outages over a year; (ii) a high-pressure, tank-fed IWS; (iii) a high-pressure, valve-controlled IWS; (iv) a high-pressure, pump-controlled IWS; and (v) a low-pressure, pump-controlled IWS. Although any change in pressure or flow can be considered a transient, we focus on and define transients as short-term pressure peaks or valleys that are outside of the range of steady-state pressures observed before and after the pressure change occurred. The study primarily focused on pressure conditions during supply start-up when transients due to pipe filling could occur.

## 2. METHODS

### 2.1. Study sites

Data were collected in five study zones experiencing a variety of IWS conditions, including four zones in Arraiján, Panama, and one zone in Delhi, India. Characteristics of each study zone, including supply continuity and how supply was controlled, are summarized in Table 1.

**Table 1** | Summary of study zones

| Zone (supply type) | Water system | Customer connections | Supply source | Average fraction of time supply was on[a] | Supply continuity |
|---|---|---|---|---|---|
| 1 (continuous, high-pressure) | Arraiján, Panama | 348 | Main transmission pipe from the treatment plant via two entrances to the zone | 99.1% | Continuous except for 11 outages during the year of monitoring and several users at high elevation |
| 2 (tank-fed, high-pressure) | Arraiján, Panama | 650 | Mostly via gravity from two storage tanks and some supply directly from main transmission pipes | 83% | Users at high elevations lost supply when storage tanks drained, which was most common during afternoons and weekends |
| 3 (valve-controlled, high-pressure) | Arraiján, Panama | 232 | Local pump station pumping directly into the distribution network, with a control valve used to direct supply to Zone 3 or other sectors | 57% | Scheduled to have 3 days with supply and 3 days without, but the actual supply varied due to irregular valve operation, pipe breaks, and pump station failures |
| 4 (pumped, high-pressure) | Arraiján, Panama | 368 | Mostly from the local pump station pumping directly to the distribution network, with small amounts of supply through small diameter pipes from other parts of the network | 87% | The pump station stopped frequently (an average of approximately once per day, but often a few times per day) due to insufficient supply or power failures, causing most of the zone to lose supply |
| 5 (pumped, low-pressure) | Delhi, India | 181 | From transmission main fed by a pump station | 22% | Supply provided for 2–3 h in the morning and 2–4 h in the evening, regulated by turning pumps on and off at the distribution reservoir |

Information on Arraiján, Panama study zones was previously reported by Erickson *et al.* (2020).

[a]Depending on the zone, this fraction was calculated based on the portion of monitoring time that pressure at the downstream monitoring station was $\geq 2$ psi at the ground level (Zones 1–3), the fraction of monitoring time that the pump station serving the zone was on (Zone 4), or the fraction of time pressure was higher than its value when supply was off (Zone 5).

### 2.1.1. Arraiján, Panama (Zones 1–4)

Arraiján is a rapidly growing peri-urban area west of Panama City, Panama. Arraiján's population grew from 60,000 inhabitants in 1990 to an estimated 263,000 in 2014 (National Institute of Census and Statistics Panama 2010a, 2010b), increasing demand for drinking water. At the time of the study, the Arraiján drinking water distribution network was supplied with an average of 585 liters per person per day, and most of the network normally had a continuous supply. Yet some areas routinely received intermittent supply due to high rates of leakage (according to utility data, only 47% of 2014 production was billed to customers, i.e., 53% non-revenue water) and because pumps, storage tanks, and/or pipes did not provide sufficient local distribution capacity. All areas of the network were subject to occasional loss of supply due to infrastructure failures such as pipe breaks, treatment plant shutdowns, and power outages affecting pump stations.

Arraiján's distribution system mainly has a branched topology, spans a large area with complex topography, and is supplied from three treatment plants. The elevation difference between the highest point and the lowest point in each of the Arraiján study zones ranged from approximately 26 to 82 m. Smaller diameter ($\leq$25 cm) distribution pipes were mostly PVC and larger diameter ($\geq$30 cm) transmission pipes were mostly ductile iron. Over half of the pipe network was <25 years old, although some portions were >35 years old.

Study Zones 1–4 in Arraiján (mapped in Supplementary Figure S2) were estimated to have between 232 and 650 connections each (not all legally registered with the utility). Pipes within these zones were PVC with diameters between 15 and 150 mm. The majority of households in intermittent areas (Zones 2, 3, and 4) stored water in their homes. A few households had larger storage tanks that filled automatically from the distribution network, but most households manually filled storage containers directly or through a flexible hose. A more detailed description of the Arraiján distribution network and supply in the Arraiján study zones is provided in Erickson *et al.* (2020).

### 2.1.2. Delhi, India (Zone 5)

Delhi is India's capital and its water utility, the *Delhi Jal Board* (DJB), serves treated piped water to 75.2% and untreated water to 6.1% of Delhi's 18 million inhabitants (3.3 million households) (Government of NCT of Delhi 2017). The DJB's water treatment capacity equates to 229 liters per inhabitant per day (Government of NCT of Delhi 2017). The DJB officially estimates their 'total distribution losses' to be on the 'order of' 40% (Government of NCT of Delhi 2017).

Study Zone 5 in Delhi covered one neighborhood (unspecified for privacy reasons) with 60 multi-story residential structures and 181 service connections. All customers had some form of water storage and most used private suction pumps to fill roof tanks (Meyer *et al.* 2021), some of which were automatically controlled. Zone 5 contained approximately 800 m of a 100–150 mm diameter cast iron pipe supplied from a 450 mm diameter cast iron trunk main, which remained at least partially full between supply cycles. The distribution network within Zone 5 had a looped topology.

## 2.2. Monitoring methods

### 2.2.1. Zones 1–4 (Arraiján, Panama)

From August 2014 to August 2015, pressure was monitored at the entrance(s) and a downstream location in each of the four Arraiján study zones (Zone 2 configuration shown in Figure 1(a) as an example), collecting between 273 and 354 days of pressure data. ECO-3 RTUs (remote telemetry units, Aquas Inc., Taipei, Taiwan) were used to monitor pressure at all locations except for the entrance to Zone 3, where pressure was monitored by an LPR-31i-200 pressure impulse recorder (Telog Instruments Inc., Victor, NY). The pressure sensors were installed in above-ground metal boxes and powered by 12-volt batteries charged with solar panels. Sensors were connected to the distribution pipe via a saddle installed on the pipe, a 15 mm PVC pipe, and a 12 mm PVC hose. Turbidity, free chlorine, and flowrate were also measured at some stations for related research (Erickson *et al.* 2017).

The ECO-3 pressure monitors normally recorded measurements every 30 s and recorded measurements every 0.1 s for a period of 2 min whenever the pressure changed more than 5% during 1 s. Specifications in the manual for the ECO-3 sensors indicated that they were rated to measure pressure down to 0 psi. However, personal communication with Aquas indicated that the sensors were capable of measuring negative pressures. Negative measurements registered by the ECO-3 sensors appeared to be qualitatively accurate based on the inspection of transient waveforms during pump start-ups. The LPR-31i-200 recorder was rated to measure pressure from −15 to 200 psi and was configured to sample pressure at 20 Hz and record the average, maximum, and minimum every 30 s. The recorder was also configured to record at 20 Hz for a period of 40 s whenever pressure changed by >10 psi within 10 s.

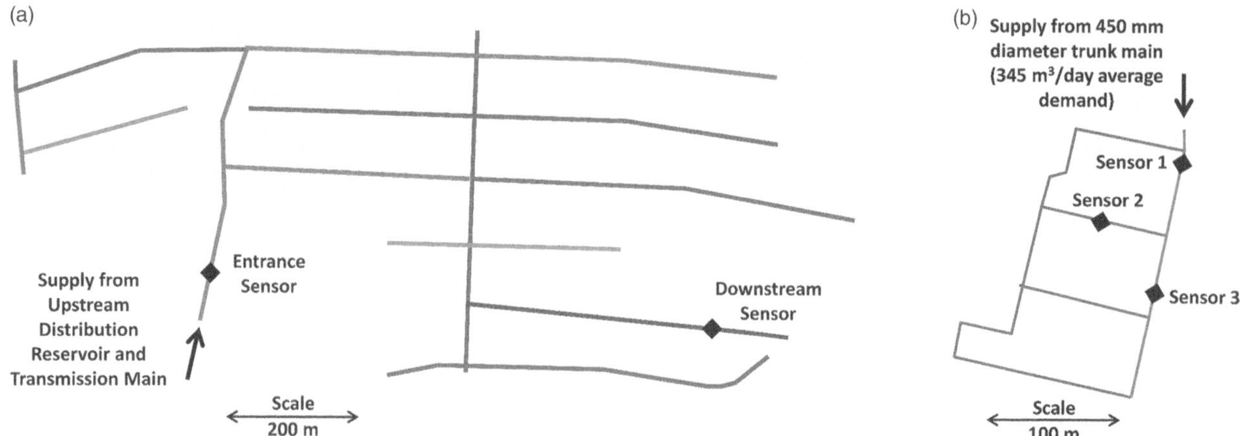



**Figure 1** | Monitoring setups in Arraiján, Panama (a; Zone 2 shown as an example) and Delhi, India (b; Zone 5). Pipe diameters were 150 mm (red), 100 mm (blue), and 75 mm (green). (a) Zone 2 (Arraiján). (b) Zone 5 (Delhi). Please refer to the online version of this paper to see this figure in colour: http://dx.doi.org/10.2166/aqua.2022.206.

### 2.2.2. Delhi, India

To observe filling pattern dynamics, three LPR-31i-100 pressure impulse recorders (same as in Arraiján, but with a maximum range of 100 psi) were installed in the neighborhood's distribution pipes (Figure 1(b)) for 2 weeks (28 filling cycles) from April 2 to 16, 2016 and configured to sample at 50 Hz and record the minimum, average, and maximum pressure every 2 min. To distinguish between full and empty pipes, pressure sensors were installed at the pipe invert (bottom of pipe).

### 2.3. Data analysis

Both datasets were analyzed and graphed in R (R Core Team 2021). For analysis of Arraiján data, transient data and periodic data (the data collected continuously every 30 s regardless of whether a transient was occurring) were combined into one time series for each monitoring location. All pressure data were adjusted to the level of the buried pipe at the monitoring location. Occasional errors in the ECO-3 sensors produced nonphysical readings, which were filtered out (filtering details in Supplementary Text S1). An algorithm was used to identify supply start-up events in the 60 min following a transition from pressure <2 psi to pressure >2 psi at ground level at the downstream monitoring station. Plots of pressure during these start-up events were then manually reviewed for patterns indicating the occurrence of hydraulic transients. Concerning transients not associated with supply start-up were identified by the manual review of pressures when the ECO-3 sensors recorded high-frequency data due to a sudden pressure change or when abnormally high or negative pressures were observed.

In Delhi, differential (gauge) pressure sensors were installed at the pipe invert, and soil saturation occasionally induced negative-pressure readings at times when customers were known to have open taps (e.g., between midnight and 2 AM). The average recorded gauge pressure (range −1, 0.4 psi) between midnight and 2 AM was used to re-zero sensors daily in postprocessing. Supply start-up events were identified based on the time the pressure became >2 psi during either the morning or the afternoon (times when the pumps were scheduled to supply the study zone). As was done with the Arraiján data, plots of pressure during these start-up events were manually reviewed for patterns indicating the occurrence of hydraulic transients.

## 3. RESULTS AND DISCUSSION

### 3.1. Downstream pressures during supply start-up

Supply start-up is a time of particular concern for transient events in IWS systems, since, if pipes have drained while the supply is off, the expulsion or collapse of entrained air pockets, and the subsequent deceleration of water columns as they run into the end of dead-end pipes (or into one another), can cause hydraulic transients. To identify potential pipe-filling transients during supply start-up, as opposed to transients from pump or valve operations, we focus our analysis on monitoring locations far downstream from pump and valve operations and close to where pipe emptying and filling were likely to be occurring. In Figure 2, pressure traces are shown from the downstream monitoring stations of Zones 1–4 (in Arraiján,

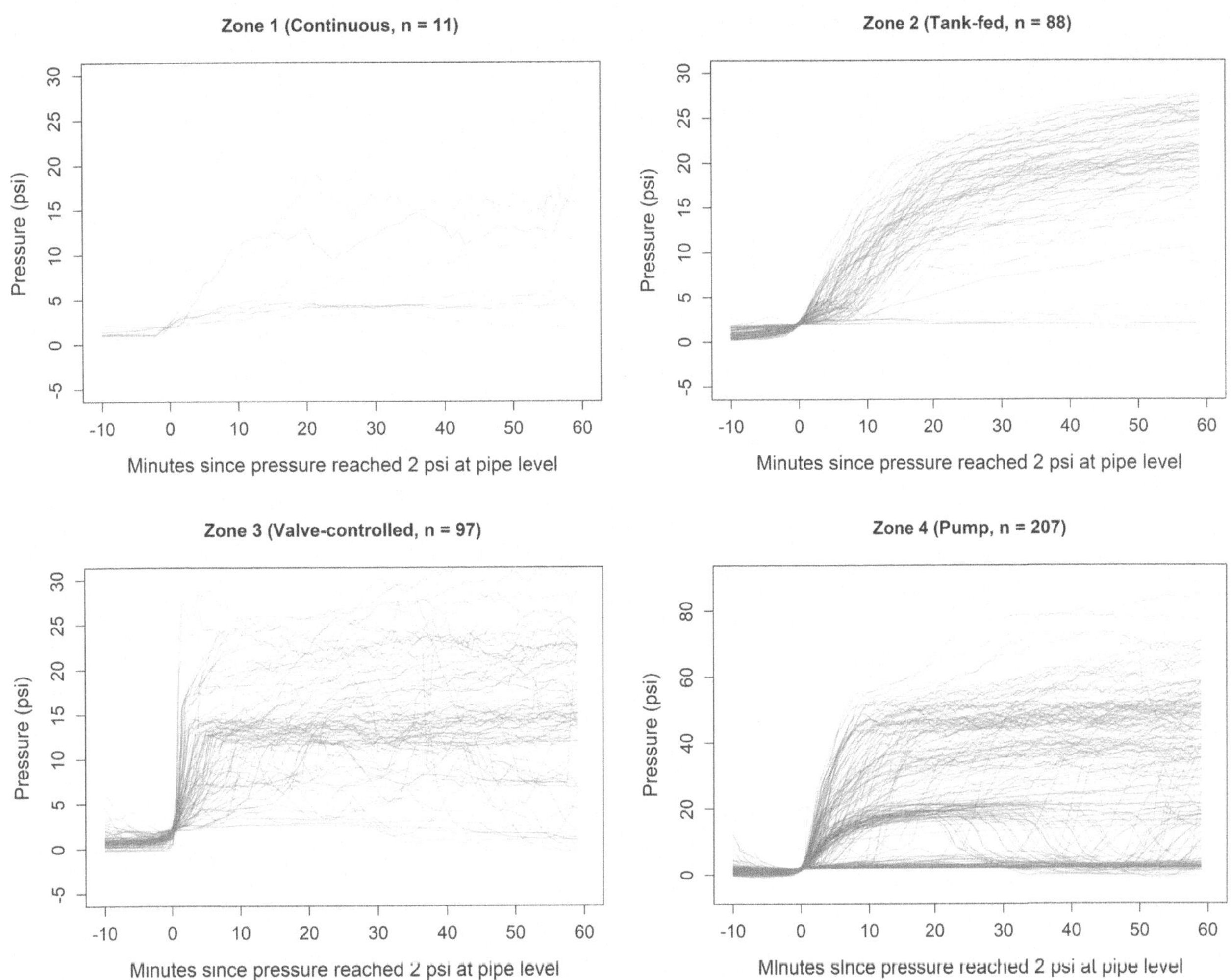



**Figure 2** | Pressure during pipe filling at downstream monitoring stations in Zones 1–4 in Arraiján, Panama. Zone 4 pressure is depicted with an expanded vertical axis.

Panama) during all supply start-up events that occurred during the year of monitoring. Each trace includes 10 min of data before and 60 min of data after pressure reached 2 psi at the pipe level.

Despite the theoretical potential for filling-induced transients, no negative or extremely high pressures were observed during the 403 supply start-up events monitored at the downstream stations. At the downstream stations in Zones 1–3, no pressures >35 psi were observed during the start-up events. The 11 filling events in Zone 1, where supply was normally continuous, were after supply outages caused by pipe breaks or maintenance. While pressures as high as 105 psi were recorded at the Zone 4 downstream monitoring station, which was supplied by intermittent pumping, these high pressures were caused by high steady-state pressures, not transients.

The downstream pressure sensors' transient detection feature was triggered for only five start-up events, all in Zone 4 (Supplementary Figure S3). While all five events were characterized by an increase in pressure when the Zone 4 pump turned on, none included the subsequent pressure oscillations typical of hydraulic transients. Pressure did oscillate during some Zone 3 and 4 start-up events (examples in Supplementary Figure S4), but the nature of these pressure changes suggests that they were not filling-induced transients.

In Zone 5 in Delhi, India, no pressure transients were observed during supply start-up (Figure 3) during the 15 morning and 15 evening filling cycles. Furthermore, for both morning and evening supply, the system appeared to reach its steady-state pressure without any substantial overshoot (Figure 3). The recorded maximum and minimum pressures during the 2-min recording intervals show trends (Supplementary Figure S5) similar to the 2-min average pressures plotted in Figure 3.

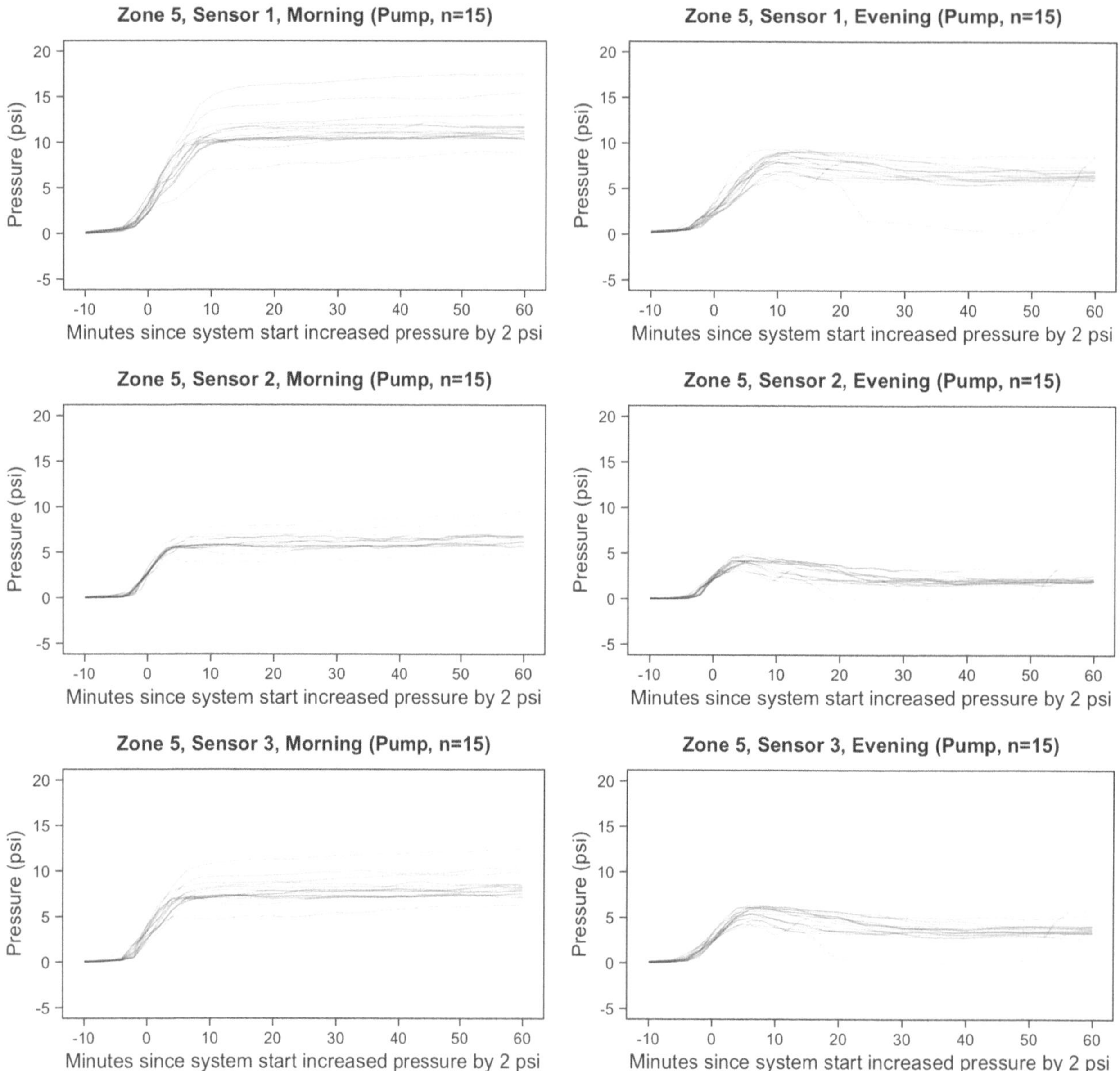

**Figure 3** | Supply start-up profiles (15 mornings and 15 evenings, measured at three sensors) in Zone 5 in Delhi, India. Pressures were recorded at 50 Hz and averaged into the 2-min recording intervals plotted. Observed max and min pressures in each recording interval are summarized in Supplementary Figure S5.

Across all five (highly varied) study zones, no substantial start-up-induced transients were observed at downstream monitoring locations, indicating that either pipes did not empty while the supply was off or that pipe filling did not result in substantial hydraulic transients.

The extent to which pipes empty while the water supply is off could affect the severity of start-up transients since there is no filling if pipes remain full during an outage. However, according to utility pipe maps, in Zones 1, 2, and 3, the downstream monitoring location was on a downhill dead-end pipe, which would presumably have been drained by customer taps further downhill on the pipe. Air was sometimes observed being expelled from customer or water quality sampling taps before supply began, suggesting that pipe draining and filling did sometimes occur.

Assuming that pipes did empty and fill, the lack of transients may be explained by the absence of a sudden transition between filling the network and the network's 'steady-state' operation. As a pipe with a downstream orifice fills, the hydraulic resistance increases when water begins to exit the orifice (instead of air), which can reduce the filling velocity and induce a pressure transient. Li & Zhu (2018) found that when the diameter of the downstream orifice was at least 17% of the pipe diameter, increasing the orifice diameter reduced the magnitude of the filling-induced pressure transient. Mimicking such downstream orifices, many customers in IWS leave taps open when the supply is off so that, when the supply returns, they will be alerted by the sound of water coming out of the tap and/or so a tank located under the tap will fill automatically when supply returns. The absence of the expected filling-associated transients in Zones 1–5 may, therefore, be explained by consumers who left their taps open during supply.

## 3.2. Concerning pressures observed in Arraiján not associated with filling

While no pipe-filling transients were observed at the downstream monitoring locations in the four Arraiján study zones, both high positive and negative transient pressures associated with other causes were observed in Arraiján, mainly at the upstream monitoring locations. These transients were associated with the irregular and complex operation that is common in intermittent systems. Uncontrolled pump starts and stops (e.g., Figure 4(a)–4(e)) and occasional pipe breaks (e.g., Figure 4(f)) caused severe pressure transients.

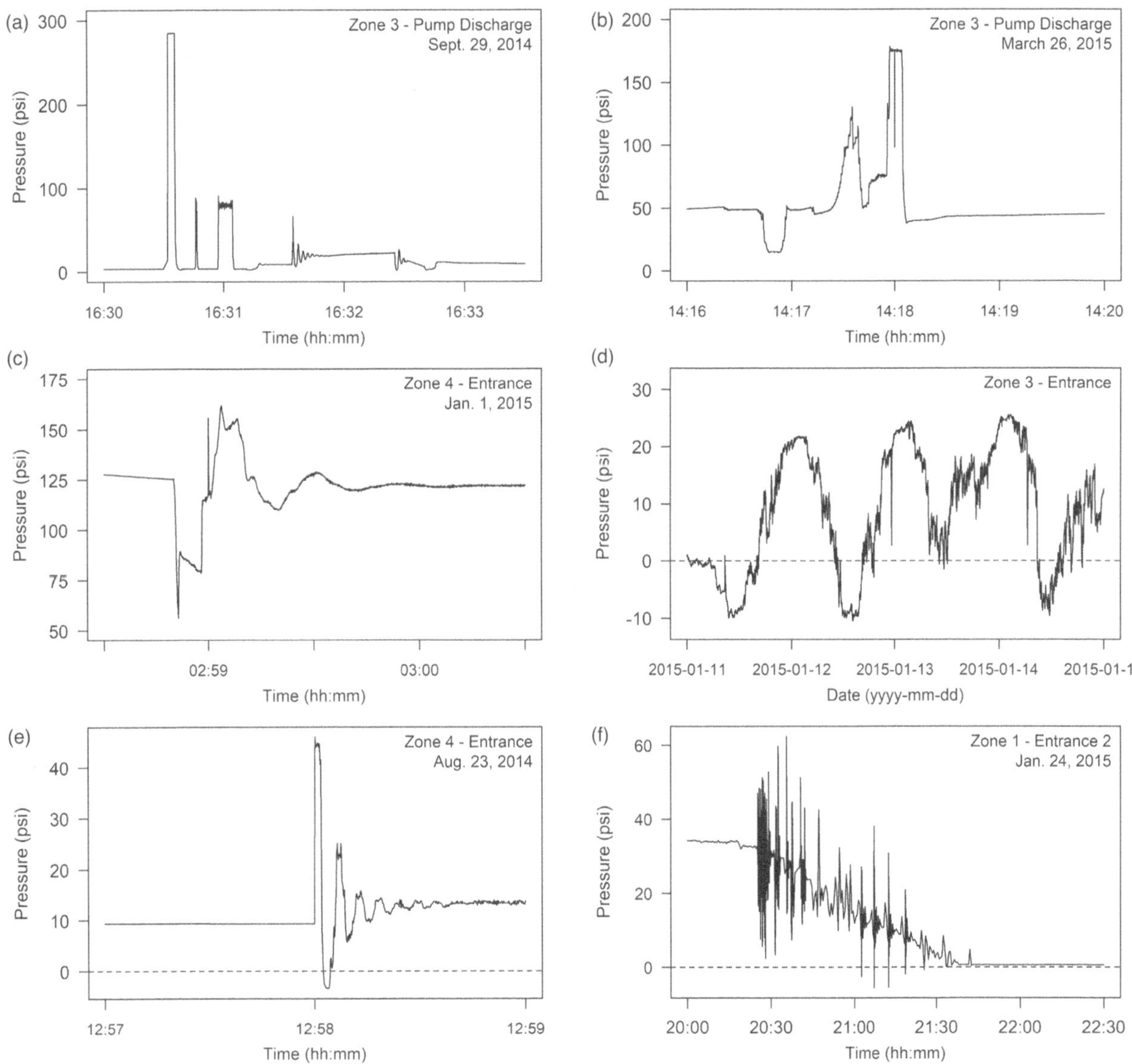

**Figure 4** | Examples of high-pressure events: (a) possible Zone 3 pump testing or repair, (b) Zone 3 pump stop and start, (c) Zone 4 pump stop and start when steady-state pressure was already high. Examples of negative-pressure events: (d) sustained negative pressures at the Zone 3 entrance, (e) Zone 4 pump start-up, and (f) transients at Zone 1 entrance from the break in a 600 mm transmission pipe.

In addition to transients, high and severely negative steady-state pressures were observed in Zones 3 and 4 (Figures 4(d) and 5(d)–5(f)). Figure 5 shows the distribution of pressures (steady-state and transient) measured at all upstream monitoring stations, where high positive and severely negative pressures were most common. Steady-state pressure at the discharge of the pump station supplying Zone 4 exceeded 132 psi during steady-state transient conditions for 0.011% of all readings, equivalent to 56 min per year (Figure 5(f)). Pressures >132 psi are above the recommended pressure limit for the DR-25 150 mm PVC pipe (AWWA 2016, based on an operating temperature of 29°C; the actual pipe class was unknown). At times, such as the pump stop and start shown in Figure 4(c), pressure transients were superimposed on already high steady-state pressures, causing further concern for pipe integrity. The pressures measured at the entrances to Zones 1 and 2 (Figure 5(a)–5(c)) and the downstream monitoring stations (see pressure distributions in Supplementary Figure S6) were more moderate.

In the zone where high steady-state pressures were most prevalent, Zone 4, the utility logged 59 breaks (Erickson *et al.* 2020) in the approximately 3 km of 150 mm PVC pipe downstream of the pump station over a 3-year period (2012–2014). While transient pressures from pump starts and stops likely played a role in the pipe breaks in this zone, the large elevation differences in the zone also contributed to extreme pressures. The Zone 4 pump station supplied areas approximately 50 m above the pump station, requiring high pumping pressures, and passed through an area approximately 30 m below the pump station, which likely experienced pressures significantly higher than those experienced at the pump station.

In Zone 3, extended, steady-state periods of negative pressure were observed (less than −2 psi for 40% of the monitoring time) at the entrance monitoring location (see Figures 4(d) and 5(e)). This location was approximately 500 m downstream of the pump station supplying the zone and at a higher elevation than the pump station on the crest of a hill over which water

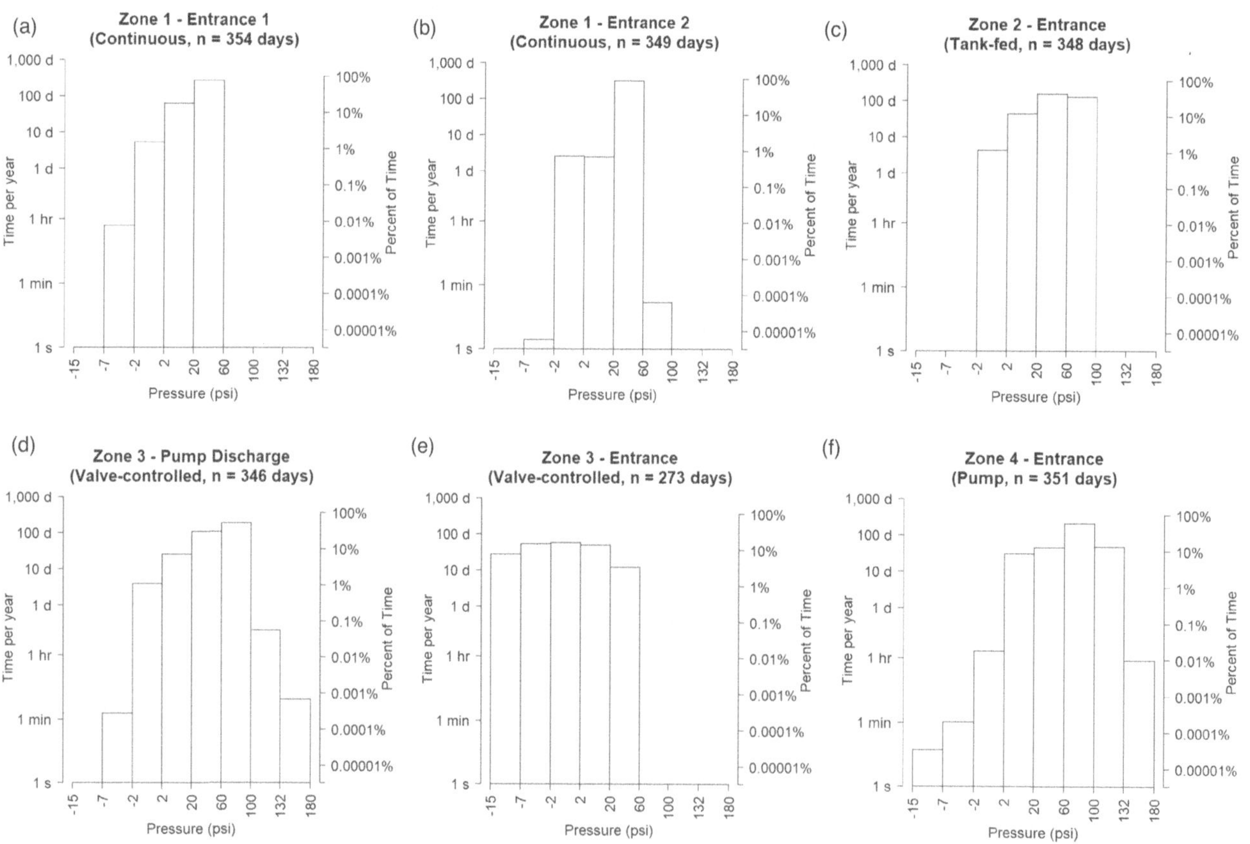

**Figure 5** | Distribution of the pressures measured at each Arraiján upstream monitoring location (Zones 1–4), including high pump station discharge pressures in Zone 3 (d) and Zone 4 (f), very negative steady-state pressures at the entrance to Zone 3 (e), and more moderate pressures at the entrances to Zones 1 and 2 (a-c). Two upstream locations are shown (a and b) for Zone 1, because water entered Zone 1 at two locations. Zone 3 upstream pressure was monitored both at the discharge of the pump station serving Zone 3 (d) and adjacent areas, and approximately 500 m downstream of the pump station, just downstream of the valve used to control supply into Zone 3 (e). The data presented include both steady-state and transient pressures. Because the number of days of data collected at each site varied, data have been normalized to the equivalent duration per year of monitoring.

was being siphoned. While PVC pipe is normally resistant to vacuum pressures (Diamond Plastics Corp 2014), negative pressures can cause contaminants to be drawn into the system (Kumpel & Nelson 2016).

Transients associated with pump operations and pipe breaks were observed in Zones 1, 3, and 4, and negative and extremely high transient and steady-state pressures were observed in Zones 3 and 4 (>100 psi for 5.6 h per year and 52 days per year at the respective discharges of Zone 3 and Zone 4 pump stations; Figure 5(d) and 5(f)). While we found no evidence of filling-induced transients, our results indicate that transients and extreme pressures can be a problem in IWS. We recommend that utilities operating intermittent (and continuous) systems leave adequate dissipation time between changes in pump operation to avoid potentially damaging transients. Monitoring for transients and analyzing pipe break data could help utilities identify and avoid operations that result in damaging pressure conditions.

## 4. CONCLUSIONS

Some aspects of IWS may aggravate hydraulic transients, while others may mitigate them. No significant pressure transients were observed at downstream monitoring locations during supply start-up in any of the five study zones in Arraiján and Delhi. However, high and negative pressures were detected in Zones 3 and 4 in Arraiján both during steady-state operation and due to or exacerbated by pump start-up and shutdown transients. Results from these field studies indicate that pipe filling in IWS does not always result in concerning pressure transients. Further research is needed to investigate why such transients do not occur in some systems and whether they do occur in other systems. We encourage future researchers working in other IWS networks to sample pressures at a high enough frequency (e.g., the methods employed in this research) to document the existence of any transient pressure conditions and to develop methods to quantify the extent to which pipes empty when supply is off. If damaging transients are more prevalent in some systems than in others, it would be beneficial to understand why so that strategies could be developed to prevent them.

## ACKNOWLEDGEMENTS

The funders gratefully acknowledge funding support from the Inter-American Development Bank (IADB, RG-T2441), a USAID Research and Innovation Fellowship, an NSF Graduate Fellowship, the Blum Center for Developing Economies and Henry Wheeler Center for Emerging and Neglected Diseases at UC Berkeley, the MIT Tata Centre for Technology and Design, and the Natural Sciences and Engineering Research Council of Canada. Additionally, the authors gratefully acknowledge the field implementation support received from J. Khari, M. Shahid, Carlos I. González, Joshua Kennedy, and Panama's National Institute of Aqueducts and Sewers (IDAAN), including Yamileth Quintero and Mauro Romero. We are also grateful for assistance from Alejandra Perroni, Stefan Buss, and Gustavo Martinez at IADB.

## DATA AVAILABILITY STATEMENT

Data cannot be made publicly available; readers should contact the corresponding author for details.

## CONFLICT OF INTEREST

The authors declare there is no conflict.

## REFERENCES

Al-Washali, T., Sharma, S., AL-Nozaily, F., Haidera, M. & Kennedy, M. 2018 Modelling the leakage rate and reduction using minimum night flow analysis in an intermittent supply system. *Water* **11** (1), 48.

Andey, S. P. & Kelkar, P. S. 2007 Performance of water distribution systems during intermittent versus continuous water supply. *Journal AWWA* **99** (8), 99–106.

AWWA 2016 *AWWA Standard C900-16 Polyvinyl Chloride (PVC) Pipe and Fabricated Fittings, 4 in. through 60 in.*

Batish, R. 2003 A New Approach to the Design of Intermittent Water Supply Networks. *World Water and Environmental Resources Congress*, Philadelphia, PA, 23-26 June 2003. ASCE, Reston, VA.

Bivins, A. W., Sumner, T., Kumpel, E., Howard, G., Cumming, O., Ross, I., Nelson, K. & Brown, J. 2017 Estimating infection risks and the global burden of diarrheal disease attributable to intermittent water supply using QMRA. *Environmental Science & Technology* **51** (13), 7542–7551.

Boulos, P. F., Karney, B. W., Wood, D. J. & Lingireddy, S. 2005 Hydraulic transient guidelines for protecting water distribution systems. *Journal AWWA* **97** (5), 111–124.

Campisano, A., Gullotta, A. & Modica, C. 2018 Using EPA-SWMM to simulate intermittent water distribution systems. *Urban Water Journal* **15** (10), 925–933.

Charalambous, B. 2011 The hidden costs of resorting to intermittent supplies. *Water* **21**, 29–30.

Christodoulou, S. & Agathokleous, A. 2012 A study on the effects of intermittent water supply on the vulnerability of urban water distribution networks. *Water Science & Technology: Water Supply* **12** (4), 523.

Coelho, S. T., James, S., Sunna, N., Abu Jaish, A. & Chatila, J. 2003 Controlling water quality in intermittent supply systems. *Water Supply* **3** (1–2), 119–125.

Cook, J., Kimuyu, P. & Whittington, D. 2016 The costs of coping with poor water supply in rural Kenya: coping costs of poor water. *Water Resources Research* **52** (2), 841–859.

De Marchis, M., Fontanazza, C. M., Freni, G., La Loggia, G., Napoli, E. & Notaro, V. 2010 A model of the filling process of an intermittent distribution network. *Urban Water Journal* **7** (6), 321–333.

Diamond Plastics Corp 2014 *PVC Pipe Under Vacuum*. Available from: http://www.dpcpipe.com/Archive/2014/4/vacuum (accessed 16 July 2016).

Erickson, J. J., Smith, C. D., Goodridge, A. & Nelson, K. L. 2017 Water quality effects of intermittent water supply in Arraiján, Panama. *Water Research* **114**, 338–350.

Erickson, J. J., Quintero, Y. C. & Nelson, K. L. 2020 Characterizing supply variability and operational challenges in an intermittent water distribution network. *Water* **12** (8), 2143.

Galaitsi, S., Russell, R., Bishara, A., Durant, J., Bogle, J. & Huber-Lee, A. 2016 Intermittent domestic water supply: a critical review and analysis of causal-consequential pathways. *Water* **8** (7), 274.

Geldreich, E. E. 1996 *Microbial Quality of Water Supply in Distribution Systems*. CRC Lewis Publishers, Boca Raton.

Ghorpade, A., Sinha, A. K. & Kalbar, P. P. 2021 Drivers for intermittent water supply in India: critical review and perspectives. *Frontiers in Water* **3**, 696630.

Government of NCT of Delhi 2017 Water supply and sewerage. In: *Economic Survey of Delhi*, Government Of NCT Of Delhi, Delhi, pp. 182–199. Available at: http://delhiplanning.nic.in/sites/default/files/ch-13.pdf.

Izquierdo, J., Fuertes, V. S., Cabrera, E., Iglesias, P. L. & Garcia-Serra, J. 1999 Pipeline start-up with entrapped air. *Journal of Hydraulic Research* **37** (5), 579–590.

Karney, B. W. & Filion, Y. R. 2003 Energy dissipation mechanisms in water distribution systems. In *ASME*. pp. 2771–2778. Available from: http://proceedings.asmedigitalcollection.asme.org/proceeding.aspx?articleid=1583443 (accessed 29 June 2016).

Karney, B. W. & McInnis, D. 1990 Transient analysis of water distribution systems. *Journal – American Water Works Association* **82** (7), 62–70.

Klingel, P. 2012 Technical causes and impacts of intermittent water distribution. *Water Science & Technology: Water Supply* **12** (4), 504.

Kumpel, E. & Nelson, K. L. 2014 Mechanisms affecting water quality in an intermittent piped water supply. *Environmental Science & Technology* **48** (5), 2766–2775.

Kumpel, E. & Nelson, K. L. 2016 Intermittent water supply: prevalence, practice, and microbial water quality. *Environmental Science & Technology* **50** (2), 542–553.

Kumpel, E., Woelfle-Erskine, C., Ray, I. & Nelson, K. L. 2017 Measuring household consumption and waste in unmetered, intermittent piped water systems. *Water Resources Research* **53** (1), 302–315.

Lee, E. J. & Schwab, K. J. 2005 Deficiencies in drinking water distribution systems in developing countries. *Journal of Water and Health* **3** (2), 109–127.

Li, L. & Zhu, D. Z. 2018 Modulation of transient pressure by an air pocket in a horizontal pipe with an end orifice. *Water Science and Technology* **77** (10), 2528–2536.

Lingireddy, S., Wood, D. J. & Zloczower, N. 2004 Pressure surges in pipeline systems resulting from air releases. *Journal AWWA* **96** (7), 88–94.

Meyer, D. D. J., Khari, J., Whittle, A. J. & Slocum, A. H. 2021 Effects of hydraulically disconnecting consumer pumps in an intermittent water supply. *Water Research X* **12**, 100107.

National Institute of Census and Statistics, Panama 2010a Cuadro 11: Superficie, población y densidad de población en la República, según provincia, comarca indígena, distrito y corregimiento: Censos de 1990 a 2010. Available from: https://www.contraloria.gob.pa/inec/archivos/P5561Cuadro%2011.pdf.

National Institute of Census and Statistics, Panama 2010b Cuadro 44: Estimación y proyección de la población del distrito de Arraiján, por corregimiento, según sexo y edad: Años 2010–20. Available from: https://www.contraloria.gob.pa/inec/archivos/P5561Cuadro%2044.pdf.

R Core Team 2021 *R: a Language and Environment for Statistical Computing*. R Foundation for Statistical Computing, Vienna, Austria.

Sánchez-Navarro, J. R., Sánchez, D. H., Navarro-Gómez, C. J. & Peraza, E. H. 2021 Multivariate analysis of the pressure variation in intermittent water supply systems and the impact on demand satisfaction. *Water Supply* **21** (7), 3932–3945.

Taylor, D. D. J., Slocum, A. H. & Whittle, A. J. 2018 Analytical scaling relations to evaluate leakage and intrusion in intermittent water supply systems. *PLoS ONE* **13** (5), e0196887.

Taylor, D. D. J., Slocum, A. H. & Whittle, A. J. 2019 Demand satisfaction as a framework for understanding intermittent water supply systems. *Water Resources Research* **55** (7), 5217–5237.

Vairavamoorthy, K., Gorantiwar, S. & Mohan, S. 2007 Intermittent water supply under water scarcity situations. *Water International* **32** (1), 121–132.

Weston, S. L., Loubser, C., Jacobs, H. E. & Speight, V. 2022 Short-term impacts of the filling transition across elevations in intermittent water supply systems. *Urban Water Journal*, 1–10.

Yepes, G., Ringskog, K. & Sarkar, S. 2001 The high cost of intermittent water supplies. *Journal of Indian Water Works Association* **33** (2), 167–170.

First received 30 June 2022; accepted in revised form 10 October 2022. Available online 20 October 2022

© 2022 The Authors

doi: 10.2166/aqua.2022.091

# Impact of intermittent supply on water meter accuracy

Marco Ferrante [a,*], Dewi Rogers [b], Josses Mugabi[c] and Francesco Casinini[a]

[a] Department of Civil and Environmental Engineering, University of Perugia, Via Duranti 93, 06125 Perugia, EU, Italy
[b] DEWI s.r.l., Via dei Ceraioli 15, 06134 Colombella, Perugia, EU, Italy
[c] World Bank, Nairobi 00100, Kenya
*Corresponding author. E-mail: marco.ferrante@unipg.it

MF, 0000-0001-5594-6555

## ABSTRACT

Supply interruptions are a common occurrence in water systems and are particularly prevalent in developing countries. During the system filling, the air that entered into the pipes when the supply was closed is expelled through any fissures including service pipes. When a meter is installed, the airflow can cause inaccuracies, over-reading, and reliability issues. This paper investigates experimentally the impact of the airflow on water meter performance during filling by means of a laboratory set-up, using pipes with diameters comparable to those of a real water system. The impact of the airflow on the accuracy and reliability of the water meters is discussed and the potential for over-reading is estimated.

**Key words**: customer meter accuracy, intermittent water supply, meter reliability, over-reading

## HIGHLIGHTS

- Supply interruptions are a common occurrence in water systems, particularly in developing countries.
- The air entering the pipes in supply interruptions is in part expelled through the customer meters causing over-reading and high rotational spin.
- Using a large experimental set-up, the paper investigates the impact of the expelled airflow on water meters.
- The estimate of over-reading and the reasons of failures are discussed.

## GRAPHICAL ABSTRACT

## INTRODUCTION

Information provided by water meters in water distribution management is relevant both for economic and hydraulic balances. For the former, the measured volumes are used to determine the customer bill while in the latter they are used to estimate leakage and non-revenue water.

It is common practice in many parts of the world to interrupt water supply when the demand exceeds the production, often leading to undesirable consequences such as pipe bursts and poor water quality (Simukonda *et al.* 2018; Taylor *et al.* 2019; Farmani *et al.* 2021). The water supply interruptions occur either by closing valves, switching off the pumps or by the natural drawdown of a tank. This action causes air to enter parts of the system which, when the supply returns, is expelled through any device or fissures, potentially including the customers' water meters. Similarly, when the supply is interrupted, air can enter pipes through the water meter (Taylor 2018).

Often the customer overcomes intermittent supply by installing storage tanks. Criminisi *et al.* (2009) have shown that, due to the inherent low flow inaccuracies over time particularly relating to the filling of tanks, water meters register less than the quantity that was delivered. The model developed by Criminisi *et al.* (2009) was implemented by Puleo *et al.* (2013) which confirmed that the cyclical emptying and filling of the tanks may cause the under-reading of the water meter. However, in the case of intermittent supply and the passage of air, this might not be the case.

The effect of the airflow through the water meters during the filling has been investigated at a laboratory scale, on small pipes. Walter *et al.* (2017) and Klingel *et al.* (2018) analysed the measurement error of single-jet (SJ) and multi-jet (MJ) water meters due to the filling of an empty pipe. The experimental set-up allowed the filling of single pipes with the same diameter of 1.95 cm and lengths ranging between 1 and 25 m, with a maximum volume of 7.72 l. The authors concluded that the measurement error, defined as the difference between the registered volume and the actual volume of water through the water meter, is mainly due to the airflow. The components due to other effects, such as air bubbles at the waterfront, unsteady flow conditions, and impact of the waterfront with the impeller, are negligible. During the tests, it was observed that dry or wet initial conditions can change the airflow velocity needed to initiate the impeller rotation because the water in the casing increases the impeller starting resistance.

As part of a wide-ranging study of the impacts of Intermittent Supply, the World Bank funded through the PPIAF programme the Department of Civil and Environmental Engineering of the University of Perugia, Italy, to undertake a research consultancy. A laboratory set-up (Ferrante *et al.* 2022) was used to investigate the impacts of pipe filling and emptying on water distribution systems. The same configuration was applied to investigate the effects of the airflow through the water meter during the pipe filling and emptying in intermittent water supply conditions. In this paper, the experimental set-up is summarised and the results of the tests are presented.

## METHODS

### Laboratory set-up

In this article, the experimental set-up consisted of a series of two polymeric pipes (Figure 1), comprising an upstream polyvinyl chloride (PVC-O) DN110 PN16 pipe and a downstream high-density polyethylene (HDPE) DN110 PN10 pipe. The same system used in Ferrante & Capponi (2017, 2018a, 2018b); Ferrante (2021) was modified to investigate the effects of filling and emptying and is described in detail in Ferrante *et al.* (2022).

The test network, composed of a series of HDPE and PVC-O pipes, with a total length of 194 m, was fed directly by the pumping system (PS), with the upstream valve PV open for the whole duration of the tests (Figure 1).

A water meter was connected to the downstream end to simulate the user connection. The most common types of water meters were used for the tests, specifically SJ and MJ water meters.

The characteristics of the water meters as defined by ISO4064-1:2017 (2017) are given in Table 1. The maximum permissible error between the minimum flow rate $Q_1$ and the transitional flow rate $Q_2$ ($\varepsilon_{1-2}$) is $\pm 5\%$ and the maximum permissible error between $Q_2$ and the overload flow rate $Q_4$ ($\varepsilon_{2-4}$) is $\pm 2\%$. By definition, $Q_4$ is the highest flow rate at which the water meter is to operate for a short period of time while the permanent flow rate $Q_3$ is the highest flow within the rated operating conditions at which the meter is to operate within the maximum permissible error. As stated by ISO4064-1:2017 (2017), water meters are supposed to measure the volume of water passing through them and so the characteristic values of discharge and errors in Table 1 do not consider airflow.

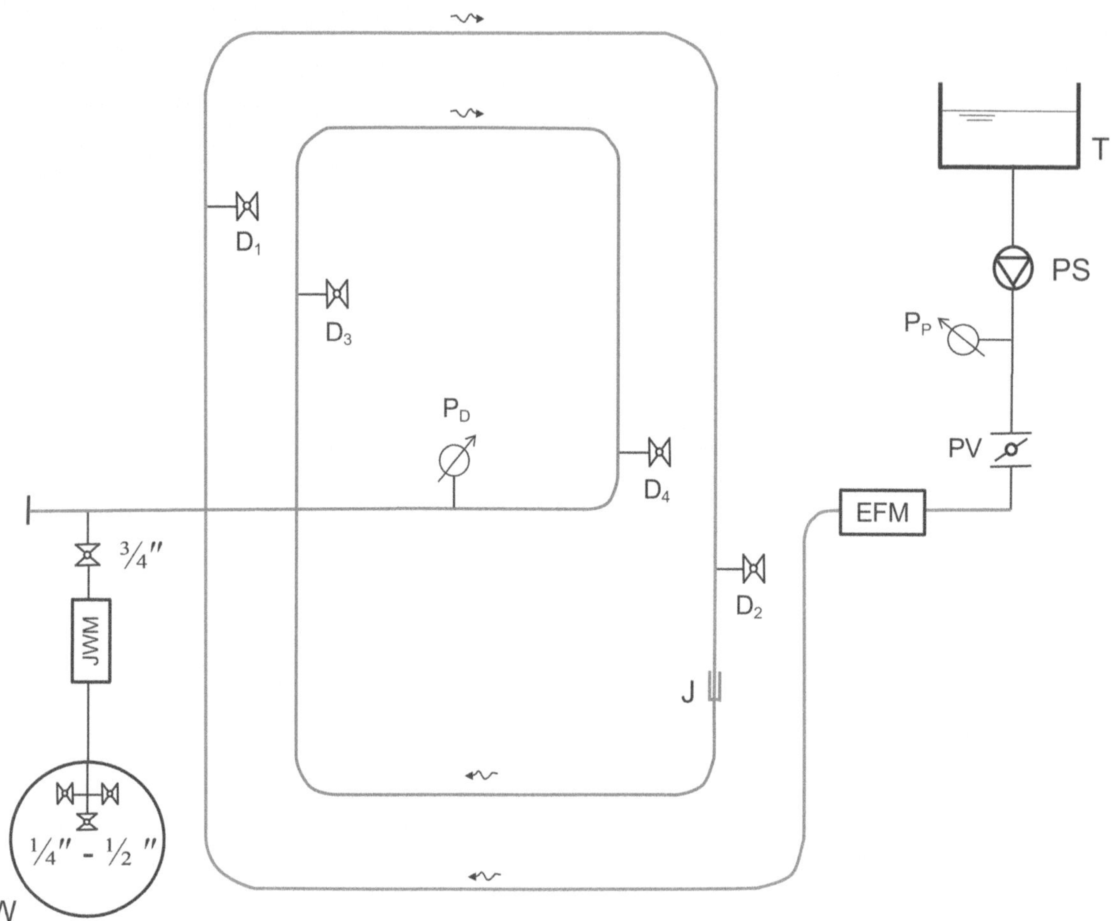

**Figure 1** | Schematic of the laboratory set-up, after Ferrante *et al.* (2022). The system is fed by the PS connected to the upstream tank T. The upstream valve PV was always open during the tests. J is the spigot junction between the upstream PVC-O pipe and the downstream HDPE pipe. $P_n$ and $D_n$ are pressure transducers and drains, respectively. The EFM is the electromagnetic flow meter. PV is a butterfly valve. The JWM is connected to the pipe by a 3/4″ ball valve. The water flowing through the JWM is collected in tank W. Three 1/4″ valves or one 1/2″ valve can be used to vary the discharge through the water meter.

The jet water meter (JWM) was connected to the end of the pipe by a 3/4″ valve, with straight steel DN20 connections longer than five diameters placed upstream and downstream, as required by good practice, to optimize the operating conditions of the meter. Downstream of the JWM and of the steel connection, the outflowing water was delivered by a DN25 HDPE pipe with a 2 m length and collected in a tank (W). Ball valves of differing diameters were installed at the downstream end of this pipe to regulate the water discharge in W. In the shown tests, one 1/4″ ball valve, one 1/2″ ball valve or three 1/4″ ball valves in parallel were used.

Two pressure transducers were used during the tests. A TRAFAG ECT 8473 absolute pressure transducer ($P_P$), with a 10-bar full scale (f.s.) and an accuracy of 0.30% f.s., was connected to the manifold downstream of the PS. A GE UNIK 5000 relative pressure transducer ($P_D$), with 6 bar f.s. and an accuracy of 0.1% f.s., was connected 2 m upstream of the pipe end.

**Table 1** | Characteristics of the water meters

| | | DN | R | $Q_{start}$ (m³/h) [(l/s)] | $Q_1$ (m³/h) [(l/s)] | $Q_2$ (m³/h) [(l/s)] | $Q_3$ (m³/h) [(l/s)] | $Q_4$ (m³/h) [(l/s)] | $\epsilon_{1-2}$ | $\epsilon_{2-4}$ |
|---|---|---|---|---|---|---|---|---|---|---|
| SJ | Single-jet | 15 | 160 | $4-5 \times 10^{-3}$ | $15.63 \times 10^{-3}$ | $25.01 \times 10^{-3}$ | $2.5\ [6.9 \times 10^{-1}]$ | $3.13\ [8.69 \times 10^{-1}]$ | $\pm 5\%$ | $\pm 2\%$ |
| MJ | Multi-jet | | | $[1.1-1.4 \times 10^{-3}]$ | $[4.342 \times 10^{-3}]$ | $[6.947 \times 10^{-3}]$ | | | | |

*Note*: $Q_{start}$ is the starting flow rate. $Q_1$ is the minimum flow rate. $Q_2$ is the transitional flow rate. $Q_3$ is the permanent flow rate. $Q_4$ is the overload flow rate. $\epsilon_{1-2}$ is the maximum permissible error between $Q_1$ and $Q_2$ and $\epsilon_{2-4}$ is the maximum permissible error between $Q_2$ and $Q_4$ (ISO4064-1:2017 2017).

An ISOIL Isomag 2500 electromagnetic flow meter (EFM) was used to measure the discharge in steady-state conditions, having an accuracy of 0.2% of the measured value. The flow meter can measure the mean flow velocity only when the pipe is completely filled with water.

A pulse unit comprising an electrical switch operated by a magnetic field (reed switch) was connected to one of the counter hands of the JWM, generating one pulse per litre.

The data acquisition system was based on a National Instruments Compact-DAQ NI-9188 chassis, which provided signal conditioning and analogue-to-digital conversion. For the purposes of the research activity, a graphical interface in LabView was developed to acquire synchronised data from the C-DAQ and evaluate the JWM flow rate by the pulses sent from the pulse unit.

The acquisition frequency of the data was set to 1 Hz for the EFM and 2,048 Hz for the pressure transducers. Details about the characteristics of instruments and data acquisition systems are also given in Ferrante *et al.* (2022).

## Tests

The results of six tests are shown in the following sections, obtained as the combination of two types of water meters, SJ or MJ, with three different downstream conditions, i.e. one 1/4″ valve, three 1/4″ valves in parallel or one 1/2″ valve open downstream of the water meter.

The aims of the tests were to investigate the impact of the airflow on the water meter reliability and accuracy, and on the over-reading.

All tests started in the empty pipe conditions with the switch-on of PS. The water entering the pipe from the upstream cross-section caused the propagation of a waterfront in the pipe and the consequent expulsion through the JWM and the downstream open valve(s) of the air beyond the front. The observed water-filling process is usually modelled as a rigid water column propagation because the velocity of the water varies in time but not in space.

Due to the combination of downstream valve air outflow and upstream water inflow conditions, the air pressure increased during the filling. Once the filling was complete and the waterfront reached the downstream end of the pipe, a sudden increase in the water pressure was observed at the two transducers, typical of a water hammer phenomenon. This increase can be explained by the deceleration of the flow caused by the difference between the discharge associated with the waterfront propagation at the arrival and the achievable water outflow discharge through the downstream valve.

After the complete filling of the pipe and the establishment of steady-state conditions, the PS was switched off. Drains $D_1$–$D_4$ were then opened to allow the pipe emptying and the start of the following test.

## RESULTS

Figure 2 shows the pressure head ($h$) variations in time ($t$) measured during the tests at $P_P$ ($a$, $b$ and $c$) and $P_D$ ($d$, $e$ and $f$), for one 1/4″ valve ($a$ and $d$), three 1/4″ valves ($b$ and $e$) and one 1/2″ valve ($c$ and $f$) open downstream of the JWM. During the pipe filling, the transducer at $P_P$ measures the water pressure at the upstream end of the water column propagating in the pipe while the transducer at $P_D$ measures the air pressure downstream of the waterfront.

The first pressure variation at $P_P$ in Figure 2(a)–2(c) with a peak at about 10 s corresponds to the pump switch on and the beginning of the pipe filling (hollow diamond).

In the same figures, the pressure peaks after the filled diamond marker correspond to the arrival of the waterfront at the downstream end of the pipe and hence the condition of a fully filled pipe.

After the pump switched on and before the arrival of the waterfront, the pressure increase at $P_D$ corresponds to the increase of the air pressure in the pipe. In fact, the air pressure can be considered to vary in time and not in space in the whole mass of air.

As previously mentioned, the air pressure variation depends on the waterfront propagation speed, which causes a reduction of the air volume and hence the air compression, and on the dimensions of the downstream valves, which cause a reduction of the air mass in the pipe and hence a pressure decrease.

Figure 3 provides a 'dual' description of the same phenomenon as shown in Figure 2, in terms of discharges ($q$), showing both the discharge variations measured at the EFM and JWM.

Although the EFM is located close to the upstream end of the pipe, at the beginning of the pipe filling and sometimes after it, it is not able to measure the water flow probably because the pipe cross-section is not yet completely full of water, the presence of air bubbles or other disturbances. Data exceeding the scale limits in Figure 3 correspond to measurement errors.

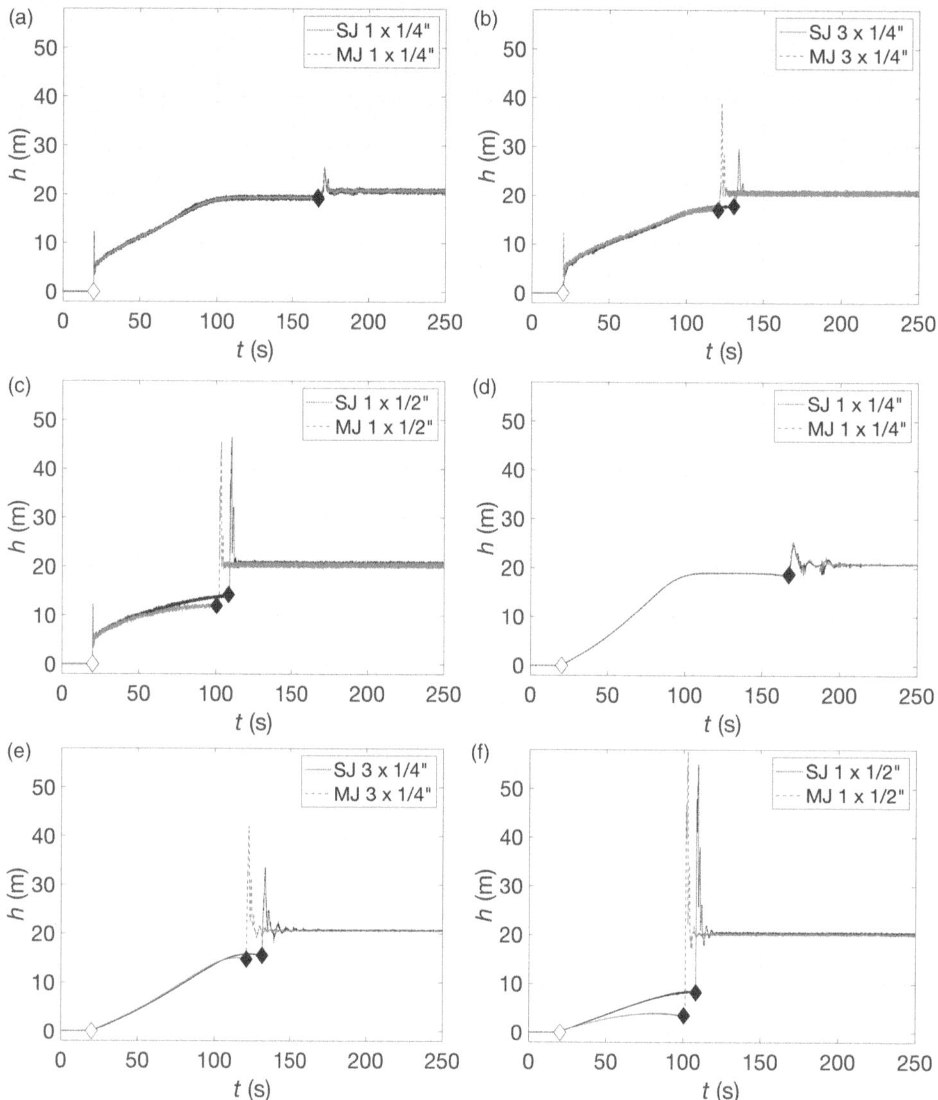

**Figure 2** | Variation in time ($t$) of the pressure head ($h$) at (*a, b* and *c*) $P_P$ and (*d, e* and *f*) $P_D$ during filling tests with (*a* and *c*) one 1/4″ valve, (*b* and *e*) three 1/4″ valves and (*c* and *f*) one 1/2″ valve, open downstream of the JWM, with SJ and MJ water meters. Hollow and filled diamonds denote the pump switch on and the arrival of the waterfront at the downstream end of the pipe, respectively.

While $P_D$ measures the air variation in time not depending on space, the EFM measures the time variation of the water rigid column discharge, which does not vary in space. The JWM measurements are linked to air discharge, although a discussion of such a relationship is given in the following sections. As a general remark, there was no evidence of a negative registration of the JWM due to the air inlet after the pump switched off. For the sake of comparison, measurements of the JWM during the tests are also shown all together in Figure 4.

Flow and water pressure at the upstream end (flow meter data in Figure 3(a)–3(c) and pressure head in Figure 2(a)–2(c), respectively) are linked by the pump characteristic curve while air and water pressures vary depending on the water column propagation and on the downstream valve opening. More detail about the pressure variations at the measurement sections during the filling is given in Ferrante *et al.* (2022). The discussion of the dependence of pressure peaks and pressure head variations with the downstream end conditions during the filling is beyond the scope of this paper.

Some differences can be seen in the pressure and discharge variations between SJ and MJ tests, especially when the three 1/4″ and one 1/2″ valves are used. As will be described later, the differences between the results of the tests with the 1/2″ downstream valve can be explained by the failure of the multi-JWM.

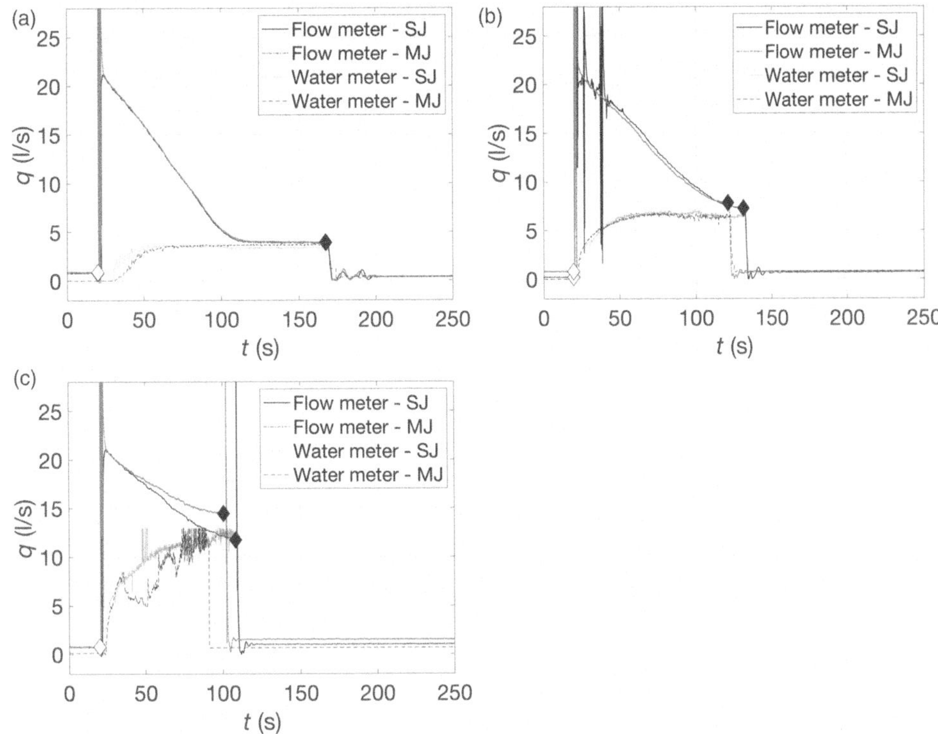

**Figure 3** | Variation in time ($t$) of the discharge ($q$) at the flow meter EFM and at the SJ and MJ water meters during filling tests with (*a*) one 1/4″ valve, (*b*) three 1/4″ valves and (*c*) one 1/2″ valve open downstream. Hollow and filled diamonds denote the pump switch on and the arrival of the waterfront at the downstream end of the pipe, respectively. Data exceeding the $q$-axis limits are due to measurement errors.

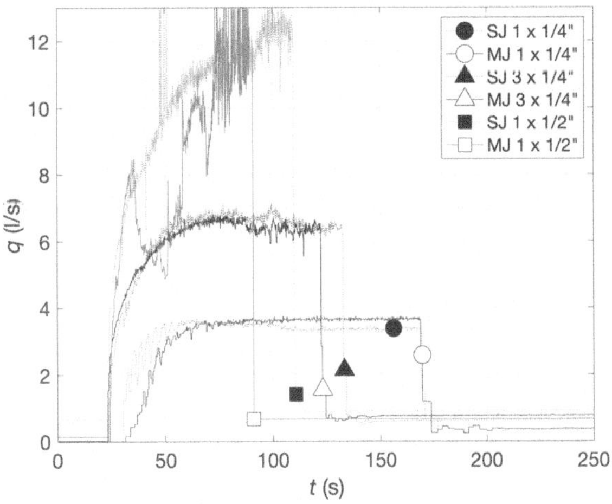

**Figure 4** | Discharge measured by the water meters during the tests with SJ or MJ water meters, for one 1/4″ valve, three 1/4″ valves or one 1/2″ valve open downstream of the water meter. Markers are used to help the line identification.

## ANALYSIS AND DISCUSSION

### The airflow through the water meters

Since the filled diamond markers in Figures 2 and 3 define the waterfront arrival at $P_D$, the JWM measurement until the marker time does not correspond to water flow but to air. The constant level of tank W confirmed that no water was collected during this period of time. On the contrary, the EFM measures the upstream water discharge that, in the rigid column propagation hypotheses, is constant along the pipe. After the transient caused by the waterfront arrival at the downstream valve, when steady-state conditions are reached, both JWM and EFM measure the water discharge.

Assuming that both air and water density were constant during the tests and that the water meter was actually able to measure the air discharge, the volume inflow measured by the EFM should be equal to the volume outflow measured by the water meter. As a consequence, the discharge measured by the JWM and EFM should be the same, which is not what the results in Figure 3 show.

Differences in measured air and water discharges cause differences in the evaluation of air and water volumes. As a matter of fact, considering the pipe lengths and inner diameters, the volume of the empty pipe is about 1.50 m$^3$. This value can be compared with the volume of the incoming water evaluated by the integral of the discharge measured by the EFM during the pipe filling in the shown tests ($V_w$) ranging between 1.37 and 1.48 m$^3$ (see Table 2). The differences between the pipe volume and the water volume entered in the pipe during the filling cannot be explained by the EFM accuracy, which could justify differences of a few litres. A possible explanation could be the unreliability of the EFM measurements at the beginning of the filling. In fact, immediately after the pump is switched on, the presence of air causes data to be missed when the water discharge reaches the highest values (see Figure 3). The missing data correspond to unaccounted-for water volumes. Another explanation could be a different amount of water in the pipe at the beginning of each test, although the time lag between two successive tests of about 1 h, with the opening of four drains along the pipe (D$_1$–D$_4$ in Figure 1), allowing the same starting conditions to be reached, which were very close to those of the empty pipe.

The outgoing air volume measured by the JWM ($V_a$) ranges between 0.445 and 0.864 m$^3$. The percentage of the air volume with respect to the water volume measured by the EFM during the same test, which provides an estimate of the empty volume at the beginning of the test, varies between 32.3 and 61.9% (Table 2).

## Water meter accuracy and airflow

The volume measured during the pipe-filling phase is due to airflow and causes over-registration. One of the objectives of the research activity was to evaluate this quantity. Hence, an issue to be addressed to relate the empty pipe air volume and the over-registration is the accuracy of the meter to measure air volume and mass. With reference to the functioning conditions during the filling, the tests show that the measured air discharge is almost always greater than the water meter overload flow rate $Q_4$, which defines the upper limit of a water meter's assured accuracy. Results in Table 2 show that the maximum value of air discharge measured during the tests ($q_{a\_max}$) ranges between 424.5 and 1,403.2% in $Q_4$.

For the same tests, it can be questioned if the valves can be considered representative of the user demand and hence of the normal functioning conditions of the water meter. To this aim, the mean values of discharges after the arrival time of the waterfront ($q_{wm}$) evaluated in filled pipe conditions for $200 \leq t \leq 250$ s, are shown in Table 2. The obtained values are compared with the maximum discharge value associated by the standards to the water meter functioning conditions, i.e. $Q_3$ and $Q_4$ in Table 1:

- for tests with the 1/4″ downstream valve open, $q_{wm} \sim Q_3/2$;
- for tests with three 1/4″ downstream valves open, $q_{wm} \sim Q_3$;
- for tests with the 1/2″ downstream valve open, $q_{wm} > Q_4$.

**Table 2** | Summary of some of the results of the test

| | SJ-1/4″ | MJ-1/4″ | SJ-3 × 1/4″ | MJ-3 × 1/4″ | SJ-1/2″ | MJ-1/2″ |
|---|---|---|---|---|---|---|
| $V_w$ (m$^3$) | 1.368 | 1.485 | 1.395 | 1.377 | 1.397 | 1.395 |
| $V_a$ (m$^3$) | 0.452 | 0.655 | 0.864 | 0.445 | 0.580 | 0.558 |
| $V_a/V_w$ (%) | 33.1 | 44.1 | 61.9 | 32.3 | 41.5 | 40.0 |
| $q_{a\_max}$ (l/s) | 3.690 | 3.743 | 7.169 | 6.797 | 12.30 | 11.76 |
| $q_{a\_max}/Q_4$ (%)[a] | 424.5 | 430.5 | 824.6 | 781.7 | 1,403.2 | 1,352.6 |
| $q_{wm}$ (l/s) | 0.335 | 0.363 | 0.648 | 0.774 | 0.926 | 0.673 |

*Note*: SJ and MJ denote single-jet and multi-jet water meters, respectively. One 1/4″ valve, three 1/4″ valves or one 1/2″ valve, were used during the tests. $V_w$ is the volume of the incoming water during the filling measured by the EFM. $V_a$ is the volume of the outgoing air during the filling measured by the JWM. $q_{a\_max}$ is the maximum discharge value measured during the test by the JWM. $q_{wm}$ is the discharge measured after the pipe filling by the JWM. The characteristic values $Q_3$ and $Q_4$ of the water meters are given in Table 1.
[a]The characteristic value $Q_4$ of the water meters is given in Table 1.

These results suggest that even customer demands in the normal functioning condition range of the water meter (e.g., $q_{wm} \sim Q_3/2$) cause a rotation speed of the turbine during the filling, i.e. airflow, of more than 4 times the maximum allowed ($q_{a\_max}/Q_4 > 400\%$). In fact, the rotation speed during the pipe filling in the shown tests was as high as 14 times the rotation speed corresponding to the water flow of $Q_4$. Furthermore, during the airflow, the turbine rotates in dry conditions, which coupled with the high rotation speeds, probably contributed to the failure of five of the six water meters used in tests undertaken in the experimental research activity. This result, of operational significance to a water utility, is also demonstrated by the results of the tests with the multi-JWM and the 1/2″ downstream open valve shown in Figures 2–4 where the meter failed at around time $t = 35$ s.

Even if the meters are used in the range $Q_2 \le q \le Q_4$, it is questionable whether they provide the same maximum permissible error with air compared to water. In fact, water meters are calibrated for the measurement of volumes of water, i.e. a liquid with an almost constant value of density, in a given range of temperatures. During the measurement of air, the turbine rotation speed could depend on both air velocity and density, which vary in time during the filling.

It is worth noting that the discharge measured by the EFM during the filling decreases in time while the discharge of the JWM increases. As mentioned previously, the decrease of the input discharge can be explained considering that it depends on the rigid column velocity, which decreases in time due to the increase of downstream pressure and friction head losses. The increase in the airflow is caused by the increase in the air pressure inside the pipe. In the tests with one 1/4″ downstream valve (Figure 4(a)), the measured discharges converged approximately at the same value about 50 s before the water column arrived at the JWM. Only in these conditions, which are not typical of the other tests, results suggest that water meters provide a reliable estimate of the output air volumes in time, which coincide with the input water volumes.

No evident and significant differences were detected between the two types of meters tested.

## Air density variation

The continuity equation that can be used to relate the initial amount of air in the pipe with that measured by the water meters must be applied to the air mass and not to the air volume. The variations of the air pressure during the filling process are shown in *Errore. L'origine riferimento non è stata trovata* suggest that the density variation could explain some differences between the initial air volumes and mass and the air volumes measured by the water meters. Since there is an outflow of air mass through the downstream valve, the density variation during the test is not simply related to the pressure even if an adiabatic process of a perfect fluid is assumed. A model describing pressures, discharges and density variations during the filling is needed for this aim.

For the sake of completeness, in Figure 5 the pressure head variation in time measured at $P_D$ is plotted against the air discharge measured by the water meter. The data are consistent with the typical patterns of relationships used to evaluate the

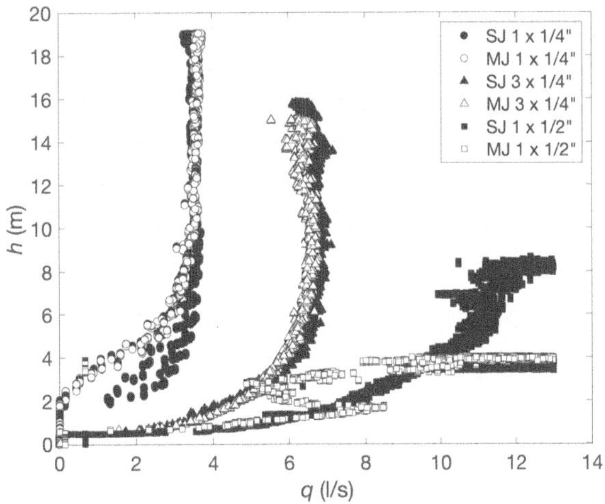

**Figure 5** | Plot of the pressure head at $P_D$ versus the discharge measured by the water meters during the tests with single- or multi-jet water meters (SJ and MJ, respectively), for one 1/4″ valve, three 1/4″ valves or one 1/2″ valve open downstream of the water meter.

airflow through an orifice (e.g., Zhou *et al.* 2020) or an air valve, with the air discharge, depending on pressures in the lower range and almost constant in the higher range of pressures (choked flow).

### Over-Reading estimation

The test results allow a rough estimate of over-reading during the pipe filling in functioning systems. In fact, the effective quantity of air measured during the filling process equated to around 0.5 m$^3$ each time the supply is interrupted. Assuming that

- the pipe diameter used in the test of 110 mm can be considered representative of a water distribution pipe in functioning systems,
- a connection density of around one property/100 m with one interruption per day,
- a mean pressure comparable to those of the tests,
- neglecting the volume of the connection to the user,

the annual over-reading could amount to as much as 90 m$^3$/year. This is likely to be an overestimate considering that there will be other outlets for the air, such as bursts and air valves. No appreciable impact was recorded of air ingress through the meter during the pipe closure which might have compensated for the over-reading.

## CONCLUSIONS

Supply interruptions are common occurrences in water systems, particularly in developing countries. This is due to the demand exceeding supply and is performed by closing the supplying valve or pumps. Customers invariably install storage tanks. An often-overlooked effect is the ingress of air when the supply is interrupted, which has to be discharged during the reopening. This air can be released through a variety of outlets, including the customer service pipes. When these are metered, it is likely that the air will have an impact on the accuracy of the registration. The present paper outlines the experimental work to evaluate the real impact of air on the measurement.

The test network of the Water Engineering Laboratory of the University of Perugia in Italy, with almost 200 m of DN110 pipes allowed the investigation of the effects of the filling of a pipe with a total volume of about 1.5 m$^3$, assessing the impact of air on the meter operation and its accuracy. The results showed the following:

- the valves downstream of the water meter used in the experiments, ranging between 1/4″ and 1/2″, as well as the water meter dimensions and the pressure close to 2 bars, can be considered representative of actual user demand conditions;
- households with installed storage tanks with open connections, to catch the supply time without manual interventions, can even have installed valves larger than those considered in the tests;
- air causes the meter to spin at around 14 times the velocity of water and around 4 times the maximum calibrated value when mean water demands are considered;
- the excessive velocity coupled with the dry state of the meter, though limited to a short time duration, can cause serious reliability issues;
- no discernible negative registration was evident due to the ingress of air at times of supply closure;
- the amount of the water meter over-registration in terms of volumes, ranging between 0.45 and 0.86 m$^3$, is significantly less than the initial air volume in the system of about 1.50 m$^3$;
- differences between air volume discharge and air mass discharge suggest that air density variation during the tests cannot be neglected;
- assuming a density of around 1 service connection per 100 m, one interruption per day would cause an over-reading equivalent to around 0.25 m$^3$/day which is probably a slight overestimate in a real network when other discharge points are also present;
- a model, which can be calibrated by the laboratory test results, is needed to improve the estimation of the over-reading in functioning systems.

In summary, the results show that the passage of air through water meters has a number of undesirable effects which ultimately will impact the customer. This relates most notably to the accuracy, over-reading and reliability of the meter. One possible solution is to use more costly meters such as electromagnetic or ultrasonic, which have no moving parts and won't read unless full of water. Alternatively, a combination check/air valve could be beneficial, even if this solution might be subject to reliability issues. Further tests and the use of a numerical model are needed in order to investigate the effects of the air density variations and the possible actions needed to mitigate the impact of intermittent supply on the water meters.

## ACKNOWLEDGEMENTS

This research was funded by the World Bank Group contract number 7199396 'An Investigation of Technological Solutions to Mitigate the Impact of Intermittent Water Supply'. The active support of Claudio Del Principe for the laboratory activities is gratefully acknowledged.

## DATA AVAILABILITY STATEMENT

Data cannot be made publicly available; readers should contact the corresponding author for details.

## CONFLICT OF INTEREST

The authors declare there is no conflict.

## REFERENCES

Criminisi, A., Fontanazza, C. M., Freni, G. & Loggia, G. L. 2009 Evaluation of the apparent losses caused by water meter under-registration in intermittent water supply. *Water Science and Technology* **60** (9), 2373–2382. https://doi.org/10.2166/wst.2009.423.

Farmani, R., Dalton, J., Charalambous, B., Lawson, E., Bunney, S. & Cotterill, S. 2021 Intermittent water supply systems and their resilience to COVID-19: IWA IWS SG survey. *Journal of Water Supply: Research and Technology-Aqua* **70** (4), 507–520. https://doi.org/10.2166/aqua.2021.009.

Ferrante, M. 2021 Transients in a series of two polymeric pipes of different materials. *Journal of Hydraulic Research* **59** (5), 810–819. https://doi.org/10.1080/00221686.2020.1844811.

Ferrante, M. & Capponi, C. 2017 Experimental characterization of PVC-O pipes for transient modeling. *Journal of Water Supply Research and Technology-Aqua* **66** (8), 606–620. https://doi.org/10.2166/aqua.2017.060.

Ferrante, M. & Capponi, C. 2018a Calibration of viscoelastic parameters by means of transients in a branched water pipeline system. *Urban Water Journal* **15** (1), 9–15. https://doi.org/10.1080/1573062x.2017.1363254.

Ferrante, M. & Capponi, C. 2018b Comparison of viscoelastic models with a different number of parameters for transient simulations. *Journal of Hydroinformatics* **20** (1), 1–17. https://doi.org/10.2166/hydro.2017.116.

Ferrante, M., Rogers, D., Casinini, F. & Mugabi, J. 2022 A laboratory set-up for the analysis of intermittent water supply: first results. *Water* **14** (6). https://doi.org/10.3390/w14060936.

ISO4064-1 2017 Water Meters For Cold Potable Water And Hot Water – Part 1: Metrological a nd Technical Requirements. International Organization for Standardization, Geneva.

Klingel, P., Mastaller, M. & Walter, D. 2018 Accuracy of single-jet and multi-jet water meters under the influence of the filling process in intermittently operated pipe networks. *Water Supply* **18** (2), 679–687. https://doi.org/10.2166/ws.2017.149.

Puleo, V., Notaro, V., La Loggia, G., Freni, G., Fontanazza, C. M. & De Marchis, M. 2013 A mathematical model to evaluate apparent losses due to meter under-registration in intermittent water distribution networks. *Water Supply* **13** (4), 914–923. https://doi.org/10.2166/ws.2013.076.

Simukonda, K., Farmani, R. & Butler, D. 2018 Intermittent water supply systems: causal factors, problems and solution options. *Urban Water Journal* **15** (5), 488–500. https://doi.org/10.1080/1573062x.2018.1483522.

Taylor, D. D. J. 2018 *Tools for Managing Intermittent Water Supplies*. Massachusetts Institute of Technology, Boston.

Taylor, D. D. J., Slocum, A. H. & Whittle, A. J. 2019 Demand satisfaction as a framework for understanding intermittent water supply systems. *Water Resources Research* **55** (7), 5217–5237. https://doi.org/10.1029/2018wr024124.

Walter, D., Mastaller, M. & Klingel, P. 2017 Accuracy of single-jet water meters during filling of the pipe network in intermittent water supply. *Urban Water Journal* **14** (10), 991–998. https://doi.org/10.1080/1573062x.2017.1301505.

Zhou, L., Cao, Y., Karney, B., Bergant, A., Tijsseling, A. S., Liu, D. & Wang, P. 2020 Expulsion of entrapped air in a rapidly filling horizontal pipe. *Journal of Hydraulic Engineering* **146** (7). https://doi.org/10.1061/(asce)hy.1943-7900.0001773.

First received 27 June 2022; accepted in revised form 19 October 2022. Available online 2 November 2022

doi: 10.2166/aqua.2023.090

# Assessing the deteriorating water quality in wards of Jaipur city through GIS interpolation

Rukshar[a], Anil Dutt Vyas [b],* and Nitu Bhatnagar[a]
[a] Department of Chemistry, Manipal University Jaipur, Jaipur, India
[b] Department of Civil Engineering, Manipal University Jaipur, Jaipur, India
*Corresponding author. E-mail: anilduttvyas@gmail.com

ADV, 0000-0003-2631-1264

## ABSTRACT

Poor management of water sources is one of the major challenge for developing nations to provide safe water to increasing population. The Central Ground Water Board, India 2019 reports that all 13 zones of Jaipur city are in the dark zone. Dark zones are those area where the ground water exploitation is extremely high and are notified for protection by the State Government. With a population of 4.1 million, Jaipur city has a water demand deficit of around 125 MLD with dependence on ground water is increasing tremendously. The city's water table has gone down 25 metres in the last decade. Higher concentrations of fluoride, nitrate, and total dissolved solids, is a common problem in the city's ground water. Presently the city gets its water from Bisalpur dam which is 120 km away, proved to be a very costly arrangement. This paper analyses the various contaminants present in ground water. A detailed survey and data collection was done, which was further analysed through GIS spatial distribution, predicting the concentration in all 91 wards of the city. Results show that some wards are at a critical point in terms of deteriorating water quality to be addressed by city planners and urban local bodies.

**Key words**: dental fluorosis, fluoride, GIS, infant mortality, nitrate, TDS

## HIGHLIGHTS

- Growing urban population in developing countries.
- Poor management of existing water sources and depletion of water table.
- High fluoride, nitrate TDS is a common ground water problem.
- Distribution of drinking water through a costly arrangement.
- GIS interpolation to ascertain various quality parameters at each ward, a guidance factor for city planners.

## INTRODUCTION

As the urban population is increasing it is having a huge impact on our natural resources including our ground water systems which are our primary sources of drinking water supply. With the change in climatic patterns, erratic and scanty rainfall, global warming has created imbalance in our ecosystems. With less rainfall, recharging of ground water sources is minimized or almost nil as artificial concrete jungles are giving less space for recharge. In almost all the towns and cities in India this is a very common phenomenon and depletion in ground water sources has aggravated the water scarcity. In a city like Jaipur the depletion of ground water is of the order of more than 3 metres every year.

Due to the depletion in ground water resources, there has been severe scarcity of water giving rise to issues of higher concentrations of total dissolved solids (TDS), arsenic, fluoride and nitrate in ground water. According to the United Nations Water report of 2017 (UN Water 2017), almost 2 billion people across the world are relying on contaminated, unsafe water sources. Almost 80% of waste water in developing nations is not treated. This untreated water contains human excreta which is going back into the ecosystem.

When compared to surface water sources, the bulk of ground water sources typically contain high amounts of fluoride. Fluoride in the soil is a result of the weathering of rocks, precipitation, and untreated water, primarily from the fertilizer industry (Ali *et al.* 2016). The majority of regular fluoride consumption comes from drinkable water (Bibi *et al.* 2017). Once a child's teeth have developed, the enamel fluorosis, which appears as spots on the enamel, never goes away (Karmakar *et al.* 2016). It is regarded as a negative consequence of fluoride interfering with ameloblasts in the growing tooth, manifesting in a disruption of the process of enamel production and increasing its porosity. The fracture risk is also increased and the bones are weakened (Bhattacharya *et al.* 2017).

Nitrate is regarded as a hazardous substance in drinking water, which is often derived from the ground. Methemoglobinemia is a common diseases where nitrate levels are too high. Blue-baby syndrome among infants is brought on by a blood condition (infant cyanosis). Infants' digestive tracts are thought to be where methemoglobin is generated when bacteria convert nitrate ions to nitrite ions (Satayeva *et al.* 2018). One nitrite molecule reacts with two hemoglobin molecules to produce methemoglobin. In an acidic atmosphere, this reaction happens quickly. The infant experiences prolonged, potentially fatal asphyxia because the changed blood protein stops blood cells from absorbing oxygen (Knobeloch *et al.* 2000). Additionally, excessive nitrate consumption can cause human disorders like Alzheimer's, vascular dementia, and multiple sclerosis (Akber *et al.* 2020).

The total amount of active charged particles dissolved in water that contains minerals, salts, and metals, is known as total dissolved solids (TDS). TDS is relevant to the quality and consistency of water as well as water filtration systems (Aher *et al.* 2017). Dissolved solids are any salts, metals, minerals, anions, or cations that may dissolve in water. The TDS concentration is determined by the quantity of anions and cations in the water (Weber-Scannell & Duffy 2007).

The main purpose of this study is to analyse the key water quality parameters such as fluoride, nitrate and TDS in Jaipur city's ground water. Detailed data collected were further analysed through GIS interpolation, and concentration of these contaminants at each ward level was determined. This is an important criterion for city planners, urban local bodies and the Public Health Engineering Department (PHED) for future city planning and zones in terms of water quality.

Till now no study has been done in Jaipur city where the concentration of major parameters has been found for each ward. This study is innovative in finding key parameters for each ward.

## METHODOLOGY

This has been done in steps.

- Study area: Jaipur City
- Jaipur drinking water scenario which includes zones, abstraction of tube wells, deteriorating ground water quality, waste water management scenario, higher concentration of various contaminants including fluoride, nitrate and TDS
- Extensive data collection in various zones
- Interpolation of data through GIS

### Study area

Rajasthan state is India's driest state, with an erratic rainfall pattern. Jaipur city, the capital city of Rajasthan, is characterized by low rainfall and high temperatures. As per Central Ground Water Board (CGWB n.d.), the annual average rainfall in the city is 504 mm, whereas the national average is 1,100 mm. The city is in the Smart City Program and is expected to have massive restructuring of urban growth. According to the Jaipur Development Authority (JDA 2020) the city had 91 wards and 8 zones. These are digitized and shown in Figure 1.

Jaipur, the 'pink city' of India and capital of the biggest state, Rajasthan, is presently facing huge water issues in terms of quality and quantity. The city has a population of 4.1 million according to the Macrotrends estimate for the year 2022 (Macrotrends 1950–2022). This is a population increase of 2.5% in comparison to 2021 and 33% in comparison to 2011. The city had 91 wards under the 2019 classifications; in this study the basis of study was confined to these 91 wards (Vyas *et al.* 2020). During 2020 the city wards were reclassified into 250 wards. With 150 wards Jaipur greater has more wards than Heritage Jaipur which has 100 wards. In this study the analysis was done for 8 zones and 91 wards as shown in Table 1.

### Jaipur drinking water scenario

Jaipur city's total water demand is more than 500 MLD. Out of this demand 275 MLD is fetched from Bisalpur dam situated 120 km from Jaipur (Jain 2020), 100 MLD is fetched from government-owned tube wells and there is demand deficit of 125 MLD. With population increasing the peak summer demand deficit presents an alarming situation. In order to cater for the remaining deficit demand there are thousands of private individual tube wells and public tube wells which are

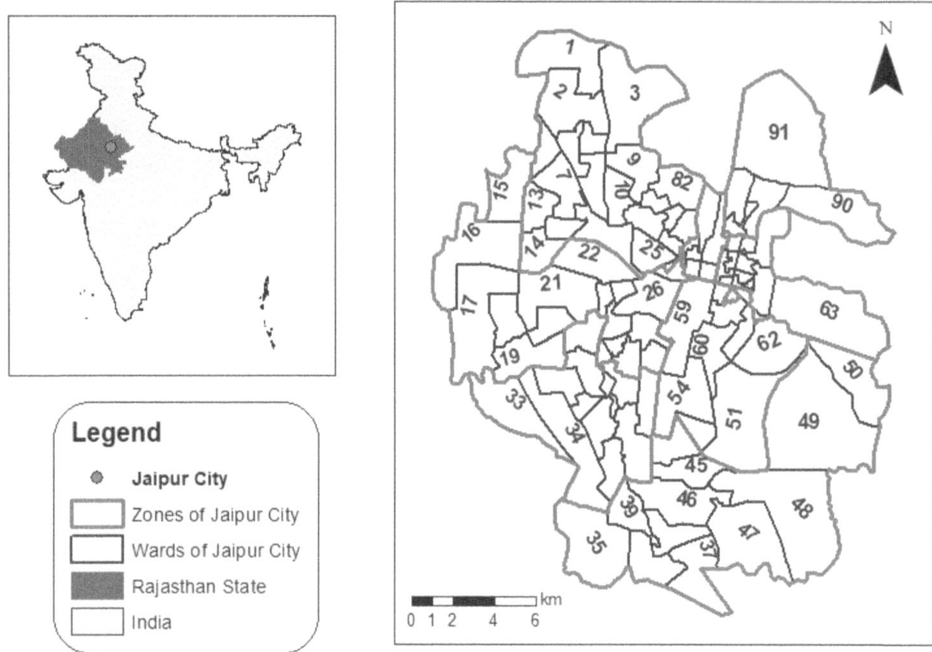



**Figure 1** | Wards and zones of Jaipur City.

supplying water to the residents through water tankers. Residents of city have been regularly complaining about the shortage and low water supply (Times of India 2018). For example, in Barkat Nagar area the situation is critical as residents receive water only for 10–15 minutes/day. The situation gets worse when many residents use boosters. On the other hand PHED reduced the water supply by 15% in the city in 2018 due to scarcity of rains. All 13 blocks of Jaipur city are overexploited and come under category dark zone according to Central Ground Water Board (CGWB 2017). A study done on Jaipur water quality by Jain *et al.* (2014) determined that more than 95% of samples have high TDS, in some cases more than 1,800 mg/l. Another study done by Roberts *et al.* (2013) predicted that there has been a declining rainfall pattern in the last two decades. Some key issues pertaining to supplying potable drinking water are outlined below.

## High abstraction from tube wells

Besides rapid urbanization in some cases ground water exploitation was done at a rate of more than 500%, one of the key factors for severe degradation of ground water quality. Illegal mining of ground water in absence of effective regulation and monitoring of drilling new tube wells have aggravated the situation. An estimated 20,000–30,000 tube wells in Jaipur city are unaccounted for (Soni 2015). Due to extreme abstraction contaminants like nitrate, fluoride, and TDS are dominant

**Table 1** | City zones and wards

| S. No. | Name of zone | Total wards |
|---|---|---|
| 1 | Vidyadhar Nagar | 21 |
| 2 | Civil Lines | 16 |
| 3 | Man Sarovar | 11 |
| 4 | Sanganer | 12 |
| 5 | Moti Doongri | 9 |
| 6 | Hawa Mahal East | 11 |
| 7 | Hawa Mahal West | 6 |
| 8 | Amer | 5 |

in the ground water. All 33 districts of Rajasthan are partially or fully affected by higher concentrations of fluoride in ground water.

### Ground water pollution

Seventy percent of Jaipur city water distribution through pipes has either high bacterial contamination or high TDS or may have both issues as projected by the India water portal based on the reports from Public Health Engineering Department Rajasthan and National Environmental Engineering Research Institute (NEERI). Jaipur district is among other districts where average concentration of fluoride is more than 2 mg/l (Singh *et al.* 2011). On the other hand, due to low rainfall, concentration of nitrate is high as ground water gets less water for dilution (Saxena & Saxena 2014). A study by Ayyasamya *et al.* (2009) states that concentration of nitrate in the Rajasthan state reaches a maximum of 1,000 mg/l. Rajput *et al.* (2019), using GIS techniques, assessed the ground water deterioration of Jaipur city. That study found that geogenic processes are more responsible for the high fluoride concentrations in ground water. Similarly excessive usage of fertilisers and domestic sewage discharge are key factors for high concentration of nitrate in the city.

### Inadequate sewage treatment plants

Research by Sharma *et al.* (2020) found that in Jaipur city only 62% of total sewage is treated through various waste water treatment plants. Only 235 MLD waste water is treated out of a total 378 MLD generation. This is one of the reasons for bacterial contamination of drinking water sources as waste water finds its way to drinking water reservoirs. On the other hand, 36% of the Jaipur city population is served by the underground sewerage system.

### Issues of fluorosis

A study done on 477 outpatients at Rajasthan dental college and hospital (Agrawal *et al.* 2014) at Jaipur revealed that only 10.9% (52) of total patients had no fluorosis. On the other hand, 89.0% (425) of total patients had fluorosis issues: 28.5% of patients (136) had moderate level of fluorosis; 30.6% of patients (146) had mild level of fluorosis symptoms; questionable level of fluorosis was 7.1% (34); and very mild level of fluorosis was 11.3% (54). Almost 58% (275) were using government supplied water as main source of drinking water and 42% (202) were relying on ground water. The severity of fluorosis is increasing at an alarming rate and the impact on school-going children is immeasurable.

### High nitrate concentration

In 2021, PHED Rajasthan chemical team collected 184 samples of water from 15 places in Jaipur (Bhaskar.com 2021). In some cases nitrate concentration was around 125 ppm against 45 ppm as prescribed by the Central Pollution Control Board. To overcome the water shortage PHED mixed the tube well water with Bisalpur dam water supply in water tankers and in clean water reservoirs (CWR) for a population of 250,000.

### Data collection: water quality parameter

Extensive data were collected for parameters related to water quality of Jaipur. These data were collected at Rajasthan state ground water department where the scientists used to collect it manually at each of the wells. Some of the data used was not primary data, i.e. it has not been collected first hand and is based on availability of data from secondary sources. Further density of data collection points for ground water quality is not optimal. A GIS database was prepared and GIS-based interpolation and zonal statistical tools were used to calculate average values at each of the wards.

In this research the focus was on fluoride, nitrate and total dissolved solids due to the deteriorating impacts on human health and their predominance in Jaipur region. The Government of India's Central Pollution Control Board (CPCB) permissible limit for fluoride is prescribed as 1.5 mg/l; this is 45 mg/l for nitrate and 500 mg/l for TDS. For this study data from 2015 to 2018 and 2021 was collected and analysed. Data for 2019 and 2020 was not available due to Covid restrictions and were not collected manually by the scientists at CGWB. For our analysis purpose we took 2015 and 2018 data. These data are shown in Tables 2–5.

The most severe sufferers from ground water quality are the urban poor who migrate to large cities for employment, education and so on, and are bound to live on the periphery or outskirts of the city area which are served water through costly individual water tankers from unsafe sources.

**Table 2** | Water quality parameters ($F^-$, $NO_3^-$ and TDS), 2015 data

| Location | Latitude | Longitude | $F^-$ (mg/L) | $NO_3$ (mg/L) | TDS (mg/L) |
| --- | --- | --- | --- | --- | --- |
| Harmara | 27.01 | 75.77 | 0.4 | 22.94 | 564 |
| Amer | 26.99 | 75.86 | 0.46 | 9.92 | 896 |
| Kukas | 27.03 | 75.89 | 0.5 | **76.88** | 465 |
| Ravinramanch | 26.91 | 75.82 | 0.6 | **182.28** | **1,076** |
| Galtagi | 26.92 | 75.86 | 0.9 | 4.96 | 164 |
| Kalwar | 26.96 | 75.68 | 1.3 | 6.82 | **554** |
| Niwaroo | 26.97 | 75.69 | 1.3 | 6.82 | 331 |
| A. Cantt. Pz | 26.98 | 75.71 | 0.4 | 9.92 | 221 |
| Jhotwara | 26.95 | 75.75 | 0.4 | 3.1 | 170 |
| Sirsi Pz | 26.91 | 75.68 | **1.9** | 19.22 | 432 |
| Niwaroo | 26.97 | 75.71 | 0.8 | 13.02 | 331 |
| Kanakpura | 26.92 | 75.72 | **1.9** | 19.84 | 313 |
| Vidhyadhar nagar | 26.97 | 75.78 | 0.6 | **63.86** | 440 |
| Govt. Hostel | 26.92 | 75.80 | 0.4 | **151.9** | **779** |
| Goner | 26.78 | 75.91 | 0.9 | 27.9 | 330 |
| Heerapura | 26.87 | 75.74 | **2.3** | **107.26** | **1,479** |
| Khori | 26.86 | 75.91 | 0.6 | **73.78** | **666** |
| Mahel | 26.81 | 75.86 | **1.9** | 40.92 | **1,144** |
| GWD Campus Pz | 26.87 | 75.82 | 0.6 | **153.76** | 595 |
| O T S | 26.87 | 75.81 | **2.3** | 21.7 | 335 |
| Bilwa | 26.76 | 75.85 | 0.9 | 40.3 | **899** |
| Muhana | 26.79 | 75.73 | 1 | 14.88 | **740** |
| Nevta | 26.80 | 75.68 | 1.3 | 14.26 | **685** |
| Sanganer | 26.82 | 75.78 | 0.3 | 14.26 | 265 |
| Dahmi kalan Pz | 26.84 | 75.57 | 0.9 | 3.1 | 391 |
| Bhankrota | 26.87 | 75.70 | 0.3 | **66.96** | **524** |

Bold values indicate more than permissible limit.

## RESULTS AND DISCUSSION

In this study three key parameters, fluoride, nitrate and TDS, pertaining to Jaipur city water quality were taken into consideration. Through the use of GIS software the concentration of all three water quality parameters was determined through spatial distribution. Through this step the concentration of fluoride, nitrate and TDS in all 91 wards for the years 2015, 2016, 2017 and 2018 was found. Based on these values the concentration of each pollutant is further shown through GIS maps in 91 wards. For the purpose of this study data for 2015 and 2018 were used. Water quality parameters of different areas of Jaipur city for 2021 were also studied.

### Fluoride ($F^-$)

Figure 2 shows fluoride concentration in all 91 wards for the year 2015. This was done through interpolation and spatial distribution. For the year 2015 all the wards of Jaipur city were well below the permissible limit of 1.5 mg/l: a healthy sign in terms of water quality pertaining to fluoride concentration.

With a similar GIS process the fluoride concentration in 2018 was higher in many zones. These included Sanganer, Hawa Mahal east area, Civil Lines area and Mansarovar area. All these areas are densely populated. This is shown in Figure 3.

Figure 4 shows fluoride concentration at various locations in Jaipur city for the year 2015. Higher concentrations of fluoride ingested for longer durations can cause skeletal problems. If ingested for a moderate duration this can lead to dental fluorosis.

**Table 3** | Water quality parameters ($F^-$, $NO_3^-$ and TDS), 2018 data

| Location | Latitude | Longitude | $F^-$ | $NO_3^-$ | TDS |
|---|---|---|---|---|---|
| Harmara | 27.01 | 75.77 | 0.58 | 11.78 | 1,098 |
| Amer | 26.99 | 75.86 | 1.02 | 104.78 | 1,209 |
| Kukas | 27.03 | 75.89 | 0.9 | 57.04 | 1,153 |
| Ravinramanch | 26.91 | 75.82 | 0.04 | 187.24 | 1,855 |
| Galtagi | 26.92 | 75.86 | 1.5 | 27.9 | 340 |
| A. Cantt. Pz | 26.98 | 75.71 | 0.8 | 96.72 | 821 |
| Jhotwara | 26.95 | 75.75 | 1.84 | 19.22 | 716 |
| Sirsi Pz | 26.91 | 75.68 | 1.48 | 19.84 | 670 |
| Niwaroo | 26.97 | 75.71 | 2.36 | 91.76 | 782 |
| Kanakpura | 26.92 | 75.72 | 2.04 | 81.84 | 727 |
| Goner | 26.78 | 75.91 | 4.2 | 70.06 | 1,835 |
| Heerapura | 26.87 | 75.74 | 2.7 | 27.9 | 569 |
| Khori | 26.86 | 75.91 | 5.56 | 70.06 | 1,555 |
| Durgapura | 26.85 | 75.79 | 0.64 | 14.88 | 341 |
| Mansarovar | 26.85 | 75.76 | 0.84 | 68.82 | 565 |
| Sanganer | 26.82 | 75.78 | 0.54 | 19.22 | 360 |
| Bhankrota | 26.87 | 75.70 | 1.16 | 13.02 | 540 |

**Table 4** | Ward-wise distribution of water quality parameters ($F^-$, $NO_3^-$ and TDS)

| Zone name | Ward No | $F^-$ 15 | $F^-$ 16 | $F^-$ 17 | $F^-$ 18 | $NO_3^-$ 15 | $NO_3^-$ 16 | $NO_3^-$ 17 | $NO_3^-$ 18 | TDS 15 | TDS 16 | TDS 17 | TDS18 |
|---|---|---|---|---|---|---|---|---|---|---|---|---|---|
| Vidyadhar Nagar | 1 | 0.7 | 1.1 | 0.6 | 0.7 | 40.3 | 26 | 59.7 | 57.0 | 426.8 | 730.7 | 458.8 | 701.5 |
| Vidyadhar Nagar | 2 | 0.7 | 0.9 | 0.6 | 0.8 | 40.3 | 26.0 | 59.7 | 57.1 | 426.7 | 732.4 | 392.7 | 700.3 |
| Vidyadhar Nagar | 3 | 0.6 | 1.0 | 0.5 | 0.5 | 40.0 | 37.5 | 59.5 | 55.8 | 423.5 | 730.7 | 580.0 | 694.2 |
| Vidyadhar Nagar | 4 | 0.7 | 0.7 | 0.6 | 0.8 | 40.3 | 30.6 | 59.7 | 57.1 | 426.7 | 741.7 | 416.8 | 673.9 |
| Vidyadhar Nagar | 5 | 0.7 | 0.6 | 0.6 | 1.0 | 40.7 | 26.4 | 57.5 | 54.2 | 410.3 | 708.8 | 362.0 | 661.7 |
| Vidyadhar Nagar | 6 | 0.8 | 0.7 | 0.7 | 1.3 | 47.3 | 21.0 | 54.6 | 51.6 | 468.6 | 657.2 | 312.9 | 640.2 |
| Vidyadhar Nagar | 7 | 0.7 | 0.4 | 0.7 | 1.3 | 46.7 | 24.7 | 53.5 | 50.2 | 452.3 | 629.3 | 346.5 | 612.6 |
| Vidyadhar Nagar | 8 | 0.7 | 0.6 | 0.6 | 0.8 | 41.6 | 31.4 | 55.2 | 51.5 | 400.4 | 676.7 | 434.1 | 634.0 |
| Vidyadhar Nagar | 9 | 0.6 | 0.9 | 0.5 | 0.3 | 40.4 | 41.1 | 58.3 | 55.1 | 424.4 | 736.7 | 593.0 | 661.4 |
| Vidyadhar Nagar | 10 | 0.6 | 0.6 | 0.6 | 0.5 | 43.6 | 37.3 | 55.7 | 50.9 | 421.4 | 693.5 | 526.6 | 620.0 |

The figure shows higher concentrations in the Jaipur centre areas which are densely populated. The fluoride concentration reached an alarming rate in two areas where the concentration was three to four times the permissible limit of 1.5 mg/l, shown in Figure 4 for the year 2018. For a few of the areas the values were not available. On the other hand 10 areas showed higher concentrations than the permissible limit.

Figure 5 displays the fluoride concentration in different parts of Jaipur for the year 2021. This was done in addition to data collected for 2015 to 2018. If more fluoride is consumed over a longer period of time, many health issues may result. Dental fluorosis may result from moderate-term consumption. The densely inhabited areas of Jaipur's centre are where the figure displays a higher concentration.

## Nitrate ($NO_3^-$)

Higher nitrate concentration is persisting in almost all the districts of Rajasthan. With the help of GIS spatial distribution interpolation as shown in Figure 6, for the year 2015 almost all eight zones of Jaipur are severely affected by nitrate except for a few wards in Vidyadhar Nagar. Not at all a healthy sign.

**Table 5** | Water quality parameters (F$^-$, NO$_3^-$ and TDS) of different areas of Jaipur city

| Rural area | Fluoride (mg/l) | TDS (mg/l) | Nitrate (mg/l) |
|---|---|---|---|
| Dehmi Kalan | 0.56 | 1,267.66 | 56.93 |
| Dehmi khurd | 0.69 | 364.00 | 27.29 |
| Sanjhariya | 0.37 | 580.00 | 16.792 |
| Awaniya | 0.25 | 384.65 | 50.898 |
| Harchandpura @ deoliya | 1.01 | 395.512 | 27.696 |
| Sitapura bas sanjhariya | 0.45 | 288.46 | 17.734 |
| Kodar | 0.71 | 423.076 | 57.162 |
| Hardhyanpura | 0.45 | 378.205 | 26.126 |
| Balmukundpura @ nada | 1.00 | 310.256 | 18.048 |
| Kumawat ki dhani | 0.39 | 392.564 | 56.968 |
| Rampurawas deoliya | 0.78 | 391.025 | 26.67 |
| Prithvisinghpura @ naiwala | 0.50 | 294.87 | 16.806 |
| Chimanpura | 0.81 | 372.00 | 25.86 |
| Bhaosinghpura | 0.15 | 537.00 | 28.132 |
| Ramsinghpura | 0.16 | 825.00 | 31.94 |
| Bindayika | 0.30 | 663.00 | 47.566 |
| Laxmipura | 0.35 | 490.50 | 30.142 |
| Himatpura | 0.32 | 487.50 | 54.764 |
| Mehlan | 0.30 | 1,110.00 | 17.75 |
| Laxminarayan pura | 0.46 | 888.50 | 38.65 |
| Girdharpura | 0.33 | 4,510.00 | 48.71 |
| Keshrisinghpura | 0.42 | 572.00 | 56.256 |
| Gopalpura | 0.42 | 609.00 | 14.26 |
| Begas | 1.18 | 340.99 | 52.75 |
| RIICO ind. area | 1.07 | 1,820.00 | 18.534 |
| Thikariya | 0.68 | 580.447 | 58.986 |
| Achanchukya | 0.56 | 391.025 | 57.144 |
| Chorasya ki dhani | 0.81 | 1,858.974 | 19.236 |
| Sarangpura | 0.24 | 576.99 | 58.516 |
| Dhami | 0.95 | 1,903.842 | 18.234 |
| Hasampura | 0.97 | 1,955.128 | 19.76 |
| Ramchandpura | 0.31 | 634.615 | 59.516 |
| Mansinghpura | 0.38 | 585.50 | 24.50 |
| Sitapura | 0.63 | 1,493.00 | 56.056 |
| Neemera | 0.35 | 780.50 | 47.112 |
| Bagru | 0.60 | 1002.00 | 58.196 |

Figure 7 shows concentration of nitrate in Jaipur region in the year 2015. Seven areas show higher concentration of nitrate. A higher concentration of NO$_3^-$ poses serious health issues mainly to infants. This has also been reported in many areas in Jaipur. A study done by Agrawal (2009) predicted that in the Dudu area in Jaipur the range of nitrate was 10–159 mg/l and almost 40% of samples showed a higher concentration of nitrate. Comparing to other years, for the year 2018 the concentration of nitrate as shown in Figure 8 was slightly lower, though in dense areas it was almost four times the permissible limit. Similarly, 10 areas have shown higher concentration than the permissible limit of 45 mg/l which is not a good healthy sign.

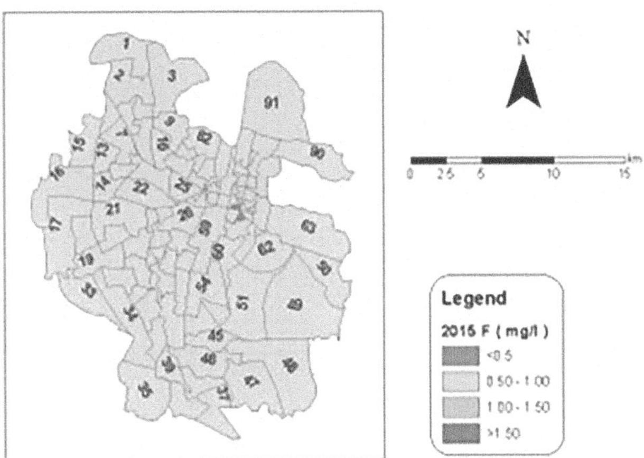

**Figure 2** | GIS interpolated F⁻ value (2015).

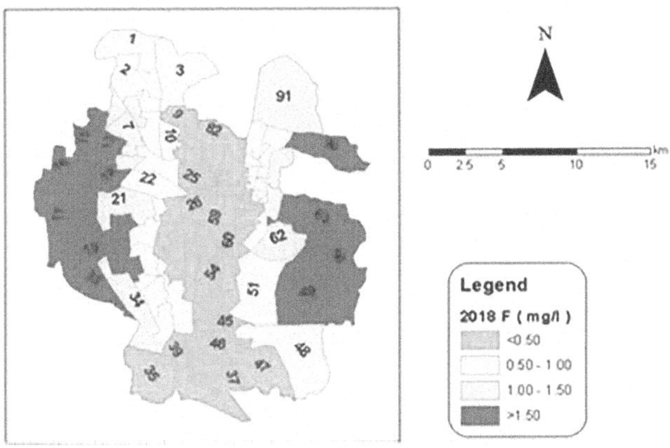

**Figure 3** | GIS interpolated F⁻ value (2018).

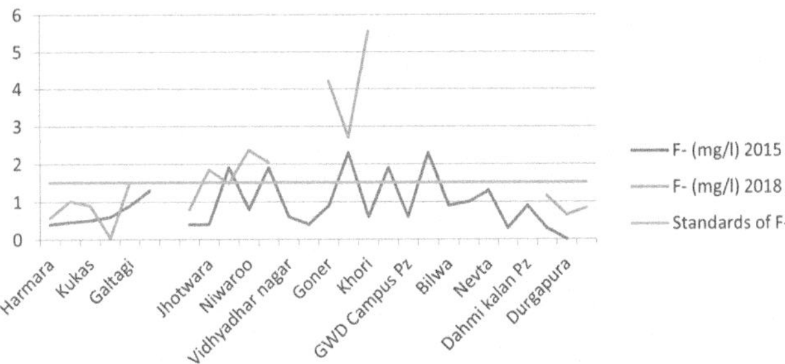

**Figure 4** | Fluoride concentration at various locations in Jaipur city for the years 2015 and 2018.

Comparing to previous years, except in Vidyadhar Nagar zone, the remaining seven zones have a better nitrate concentration in Jaipur city. This may be attributed to dilution of ground water aquifers due to rainfall. This is shown in Figure 8. Fortunately, the densely populated areas of Sanganer and Mansarovar all have better quality.

When compared to previous years, the nitrate concentration for the year 2021, as shown in Figure 9, was somewhat lower, but it was about four times higher in crowded regions. Similar to this, several locations have revealed concentrations over the allowable limit of 45 mg/l, which is not really a good indicator for health.

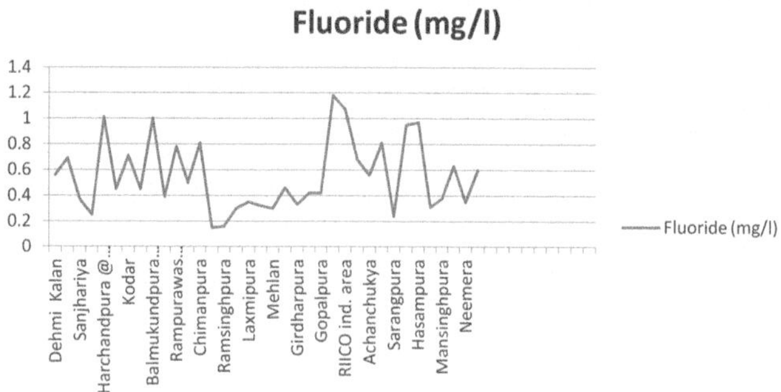


**Figure 5** | Fluoride concentration at various locations in Jaipur city for the year 2021.

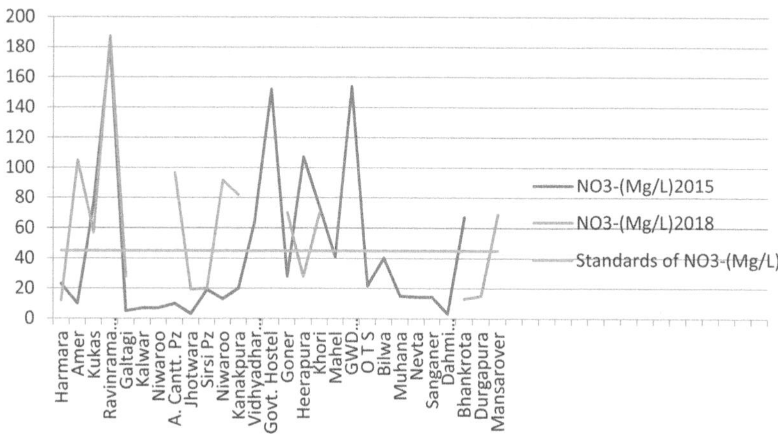


**Figure 6** | Nitrate concentration at various locations in Jaipur city for the years 2015 and 2018.

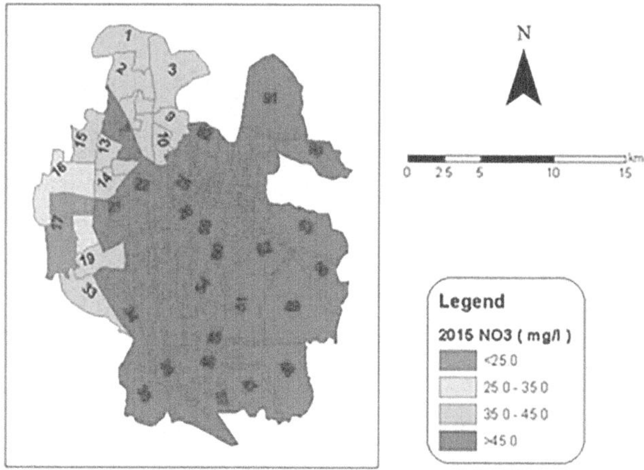


**Figure 7** | GIS interpolated $NO_3^-$ value, 2015.

## Total dissolved solids

Figure 10 shows the concentration of TDS in all 91 wards and 8 zones of Jaipur city for the year 2015. Except 2/3 wards in Sanganer zone all the wards of Jaipur have a higher TDS and 1/2 wards of Civil Lines area have concentration more than 750 mg/l. The year 2018 in contrast to the other years 2016 and 2017 shows better TDS concentration in almost all the

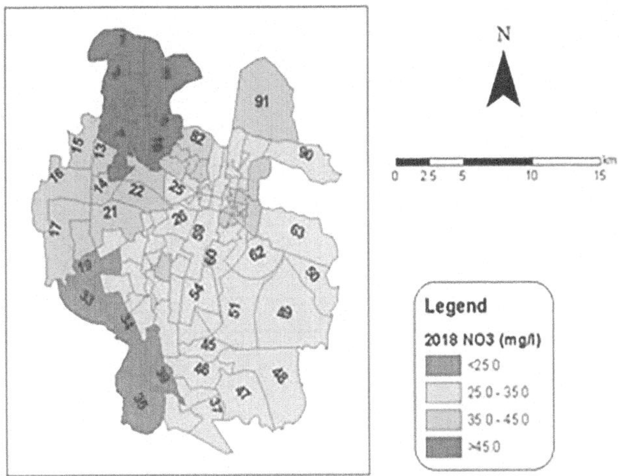



**Figure 8** | GIS interpolated $NO_3^-$ value, 2018.

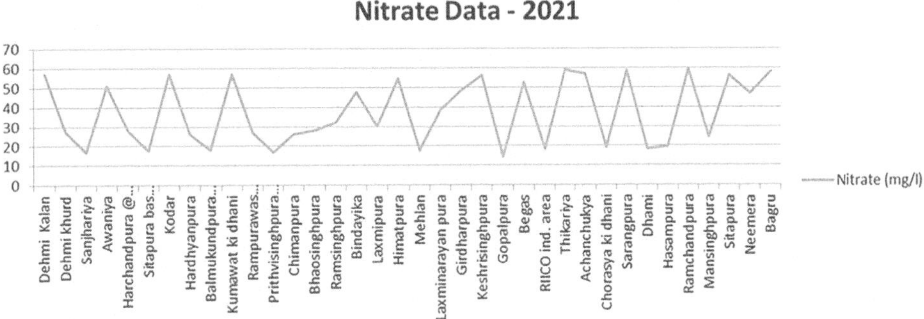



**Figure 9** | $NO_3^-$ value of water samples in different villages of Jaipur, 2021.

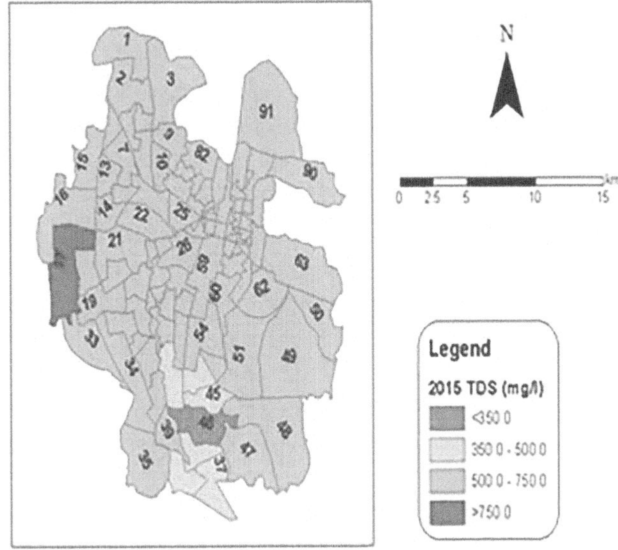



**Figure 10** | GIS interpolated TDS value, 2015.

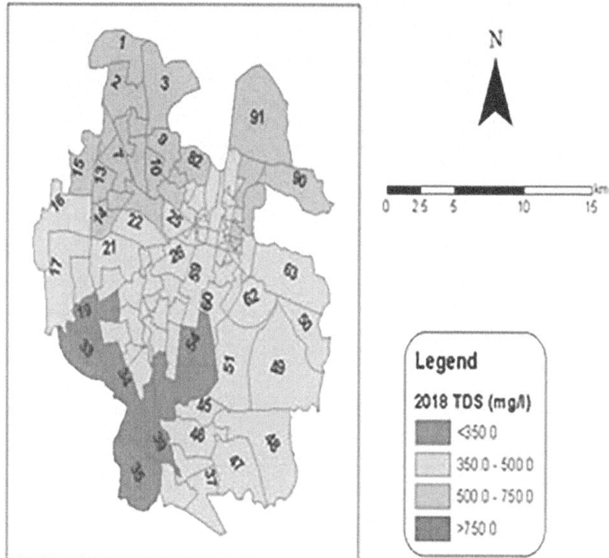

Figure 11 | GIS interpolated TDS value, 2018.

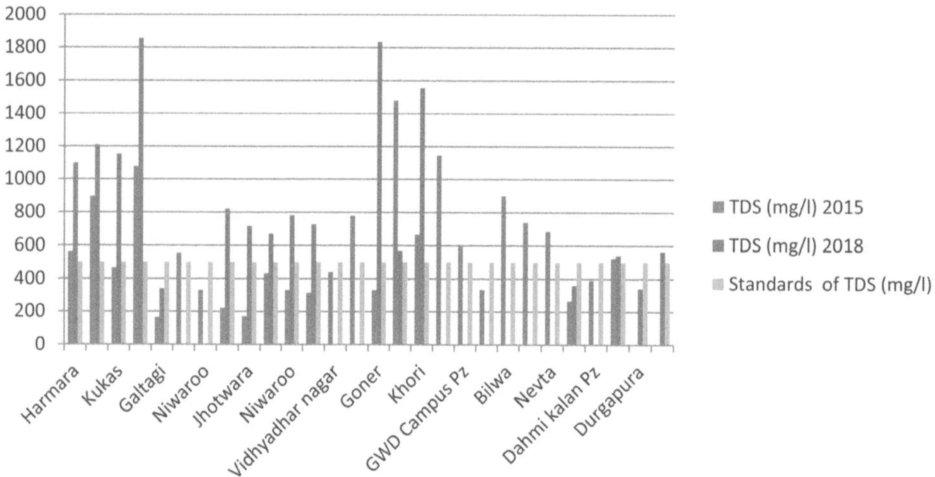



Figure 12 | TDS concentration at various locations in Jaipur city for the years 2015 and 2018.

zones and wards of Jaipur city, except the northern zones of city, that is Vidyadhar Nagar and parts of Amer. This is shown in Figure 11, and a better year in terms of TDS quality in Jaipur city.

For the year 2015 the majority of the areas in Jaipur city witnessed high TDS above 500 mg/l. Though a higher TDS does not indicate a serious health hazard, water with a concentration above 1,500/2,000 mg/l will not be acceptable by a filtration unit. Specific ions like fluoride, nitrate, and arsenic do possess serious health issues if concentration is higher than the permissible limits. Comparing to other years, for the year 2018 the TDS concentration as shown in Figure 12 in a few areas was more than 1,500 mg/l. On the other hand 15 areas showed higher concentration than the permissible limit of 500 mg/l.

The TDS concentration for the year 2021, as shown in Figure 13, was slightly lower than in previous years, due to more rainfall and recharging of ground water, but even so, it was over 1,500 mg/l in a small number of locations. On the other hand, 22 locations showed concentrations that were higher than the allowed limit of 500 mg/l.

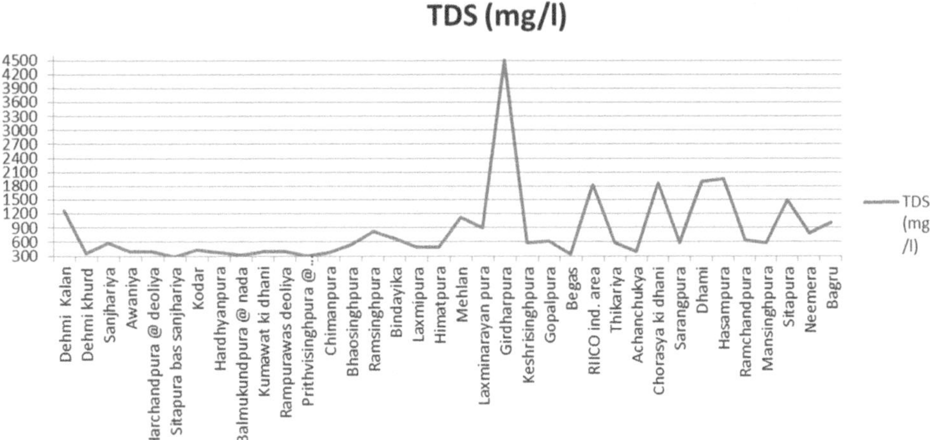


**Figure 13** | TDS concentration at various locations in Jaipur city for the year 2021.

## CONCLUSION AND FUTURE PERSPECTIVES

The most severe sufferers from ground water quality are the urban poor. They migrate to large cities for employment and education, and so on, and are bound to live on the periphery of the city. They are served water through costly individual water tankers from unsafe sources. Jaipur city has annual average rainfall of 504 mm, which is less than half of India's annual average rainfall of 1,100 mm. Climate change impact, ecological imbalance, erratic rain patterns and reduction in rainfall combined with rapid urbanization, steep increase in population and extension of concrete floors is giving much less space for ground water recharge. Without more dilution of ground water the issues of fluoride, nitrate and TDS is increasing at a very high rate. Effective rain water harvesting structures and developing green zones are two of the ways to mitigate the issue. An effective environmental management plan is urgently needed.

This study demonstrated that it is possible to compare the observed water quality parameters value favourably with the estimated water quality parameters value. As a result, the proposed model can be utilized to reasonably anticipate and manage the water quality. In order to prevent additional pollution of this major area of Jaipur city, the study advises the necessity of ongoing water monitoring in order to identify the causes of pollution.

## ACKNOWLEDGEMENTS

The authors acknowledge the kind support from various regulatory agencies in Jaipur city including Central Ground Water Board (CGWB), PHED, JDA, JMC, MNIT (Malviya National Institute of Technology), and MUJ (Manipal University Jaipur). The authors would also like to acknowledge the support of Mr Kushal Mahale, research scholar from Manipal University Jaipur with a special thanks to Dr Rohit Goyal, Professor of Civil Engineering, Malviya National Institute of Technology, Jaipur.

## DATA AVAILABILITY STATEMENT

All relevant data are included in the paper or its Supplementary Information.

## CONFLICT OF INTEREST

The authors declare there is no conflict.

## REFERENCES

Agrawal, R. 2009 Study of physico-chemical parameters of groundwater quality of Dudu town in Rajasthan. *Rasayan Journal of Chemistry* **2** (4), 969–971.

Agrawal, M., Agrawal, S., Adyanthaya, B. R., Gupta, H. L., Bhargava, N. & Rastogi, R. 2014 Prevalence and severity of dental fluorosis among patients visiting a dental college in Jaipur, Rajasthan. *Indian Journal of Research in Pharmacy and Biotechnology* **2** (4), 1339.

Aher, A., Papp, J., Colburn, A., Wan, H., Hatakeyama, E., Prakash, P., Weaver, B. & Bhattacharyya, D. 2017 Naphthenic acids removal from high TDS produced water by persulfate mediated iron oxide functionalized catalytic membrane, and by nanofiltration. *Chemical Engineering Journal* **327**, 573–583.

Akber, M., Islam, M., Dutta, M. & Billah, S. M. 2020 Nitrate contamination of water in dug wells and associated health risks of rural communities in southwest Bangladesh. *Environmental Monitoring and Assessment* **192** (3), 1–12.

Ali, S., Thakur, S. K., Sarkar, A. & Shekhar, S. 2016 Worldwide contamination of water by fluoride. *Environmental Chemistry Letters* **14** (3), 291–315.

Ayyasamya, P. M., Rajakumarb, S., Sathishkumarc, M., Swaminathanc, K., Shanthid, K. & Leea, S. 2009 Nitrate removal from synthetic medium and groundwater with aquatic macrophytes. *Desalination* **242** (1–3), 286–296.

Bhaskar.com. 2021 *Dangerous: Supply of High Nitrate Drinking Water at 15 Places Including SMS and JK Lon.* Available from: https://www.bhaskar.com/local/rajasthan/jaipur/news/supply-of-high-nitrate-drinking-water-at-15-places-including-sms-and-jk-lone-128812915.html (accessed 19 March 2022).

Bhattacharya, P., Samal, A. C., Banerjee, S., Pyne, J. & Santra, S. C. 2017 Assessment of potential health risk of fluoride consumption through rice, pulses, and vegetables in addition to consumption of fluoride-contaminated drinking water of West Bengal, India. *Environmental Science and Pollution Research* **24** (25), 20300–20314.

Bibi, S., Kamran, M. A., Sultana, J. & Farooqi, A. 2017 Occurrence and methods to remove arsenic and fluoride contamination in water. *Environmental Chemistry Letters* **15** (1), 125–149.

CGWB 2017 *Report on Aquifer Mapping and Ground Water Management: Jaipur District, Rajasthan.* CGWB, Government of India, Ministry of Water Resources. Available from: http://cgwb.gov.in/AQM/NAQUIM_REPORT/Rajasthan/Jaipur.pdf (accessed 15 March 2023).

CGWB no date *State Profile: Ground Water Scenario of Rajasthan.* Available from: http://cgwb.gov.in/gw_profiles/st_Rajasthan.htm (accessed 7 July 2020).

Jain, S. 2020 Drinking water management of Jaipur City: Issues & challenges. Web presentation. Available from: https://www.readkong.com/page/drinking-water-management-of-jaipur-city-issues-challenges-1701025 (accessed 15 March 2023).

Jain, N., Kumar, S., Lata, R., Singh, R. K., Ahmad, S. & Kumar, S. 2014 Ground water quality assessment of Jaipur City, Rajasthan (India). *International Journal of Engineering Research and Technology (IJERT), ETWQQM* **3** (03).

JDA 2020 *Master Development Plan 2025, Jaipur.* Urban Development and Housing, Government of Rajasthan. Available from: https://jda.urban.rajasthan.gov.in/content/dam/raj/udh/organizations/jaipur-nagar-nigam/pdf/Enewsletter/enewsletter-April2016.pdf (accessed 15 November 2020).

Karmakar, S., Mukherjee, J. & Mukherjee, S. 2016 Removal of fluoride contamination in water by three aquatic plants. *International Journal of Phytoremediation* **18** (3), 222–227.

Knobeloch, L., Salna, B., Hogan, A., Postle, J. & Anderson, H. 2000 Blue babies and nitrate-contaminated well water. *Environmental Health Perspectives* **108** (7), 675–678.

Macrotrends, Jaipur India Metro Area Population 1950–2022. Available from: https://www.macrotrends.net/cities/21280/jaipur/population (accessed 15 July 2021).

Rajput, H., Kumar, A. & Goyal, R. 2019 Use of improved DRASTIC model for groundwater vulnerability assessment of upper Alwar district of Rajasthan state. *ISH Journal of Hydraulic Engineering* **27** (sup1), 462–470.

Roberts, K., Reiner, M. & Gray, K. 2013 *Water Scarcity in Jaipur, Rajasthan, India.* Jal Bhagirathi Foundation. Available from: http://www.civil.northwestern.edu/EHE/HTML_KAG/Kimweb/files/Jaipur%20Water%20Resources%20(01.15.14).pdf (accessed 10 October 2021).

Satayeva, A. R., Howell, C. A., Korobeinyk, A. V., Jandosov, J., Inglezakis, V. J., Mansurov, Z. A. & Mikhalovsky, S. V. 2018 Investigation of rice husk derived activated carbon for removal of nitrate contamination from water. *Science of the Total Environment* **630**, 1237–1245.

Saxena, U. & Saxena, S. 2014 Ground water quality evaluation with special reference to fluoride and nitrate contamination in Bassi Tehsil of district Jaipur, Rajasthan, India. *International Journal of Environmental Sciences* **5** (1), 144–163.

Sharma, N., Gupta, S. & Vyas, A. D. 2020 Estimation of fuel potential of faecal sludge in a water scarce city, a case study of Jaipur Urban, India. *Water Practice and Technology* **15** (2), 506–514.

Singh, P., Rani, B., Singh, U. & Maheshwari, R. 2011 Fluoride contamination in ground water of Rajasthan and its mitigation strategies. *Journal of Pharmaceutical and Bio-Medical Sciences* **6** (6), 1–12.

Soni, N. 2015 *Water Management in Urban Rural and Slum Households in Jaipur District Rajasthan. Thesis,* University of Rajasthan. Available from: https://shodhganga.inflibnet.ac.in/bitstream/10603/148551/6/06_chapter%201.pdf (accessed 19 June 2022).

Times of India 2018 Jaipur: Water tankers turn lifeline for parched Jaipur. *Times of India,* 16 April 2018.

UN_Water_2017 *Annual_Report_2017.* Available from: https://www.unwater.org/sites/default/files/app/uploads/2018/10/UN_Water_Annual_report_2017.pdf (accessed 15 March 2023).

Vyas, A. D., Mahale, K., Ajmera, D. & Goyal, R. 2020 Optimum weights of environmental parameters in evaluating urban environmental sustainability index: case study of Jaipur city. *ISH Journal of Hydraulic Engineering* **28** (1), 62–71.

Weber-Scannell, P. K. & Duffy, L. K. 2007 Effects of total dissolved solids on aquatic organism: a review of literature and recommendation for salmonid species. *American Journal of Environmental Sciences* **3** (1), 1–6. https://doi.org/10.3844/ajessp.2007.1.6.

First received 26 June 2022; accepted in revised form 24 February 2023. Available online 24 March 2023

doi: 10.2166/aqua.2022.028

# Review of hydraulic modelling approaches for intermittent water supply systems

Dondu Sarisen a,*,†, Vasilis Koukoravas a,†, Raziyeh Farmani a, Zoran Kapelan a,b and Fayyaz Ali Memon a

a Centre for Water Systems, University of Exeter, Harrison Building, North Park Road, Exeter EX4 4QF, UK
b Department of Water Management, Delft University of Technology, Stevinweg 1, 2828CN Delft, The Netherlands
*Corresponding author. E-mail: ds573@exeter.ac.uk
†The first two authors contributed equally to this work.

DS, 0000-0002-9698-9612; VK, 0000-0003-0634-7067

## ABSTRACT

Intermittent water supply (IWS) is widely used around the world, and with the increase in population and predicted future water scarcity, IWS applications seem to continue. While most of the existing studies on water supply concentrate on continuous water supply (CWS), the research focused on the IWS is now becoming mainstream. Hydraulic modelling is an effective tool for the process of planning, design, rehabilitation, and operation of water distribution systems. It helps significantly in engineers' decision-making processes. The necessity of modelling IWS systems arises from the complexity and variety of problems caused by intermittency. This paper offers a review of the state-of-the-art IWS modelling and identifies the key strengths and limitations of the available approaches, and points at potential research directions. Currently, neither computer software nor a practically used approach is available for modelling IWS. For a rigorous simulation of IWS, system characteristics first need to be understood, i.e., the user behaviour under pressure-deficient conditions, water losses, and filling and emptying processes. Each of them requires further attention and improvement. Additionally, the necessity of real data from IWSs is stressed. Accurate modelling will lead to the development of improved measures for the problems caused by intermittency.

**Key words:** EPANET, EPA-SWMM, hydraulic modelling, intermittent water supply, macroscopic model, pressure-dependent analysis

## HIGHLIGHTS

- Reviews are provided in intermittent water supply (IWS) hydraulic modelling approaches and pressure-driven analysis methods conducted to address pressure-deficient conditions.
- The distinctive characteristics of IWS are highlighted as well as its adverse impacts, principally inequitable supply.
- Challenges of modelling IWS are identified.
- Each case study is unique, and uncertainties prevail around the practices of the IWS system.

## NOMENCLATURE

Full form of abbreviations

ADEV  Average deviation of the sum of each DN supply ratio from the ASR
ASR   Average supply ratio
CV    Check valve
CWS   Continuous water supply
DDA   Demand-dependent analysis
DN    Demand node
ECT   Elastic column theory
EOA   Equivalent orifice area
FCV   Flow control valve
GIS   Geographic information system
IWS   Intermittent water supply
MOC   Method of characteristics
NHFR  Nodal head-flow relationship

PDA     Pressure-driven analysis
PDD     Pressure-dependent demand
PRV     Pressure reduction valve
RCT     Rigid column theory
RSV     Reservoir
UC      Uniformity coefficient
WDN     Water distribution network
WDS     Water distribution system

## 1. INTRODUCTION

Intermittent water supply (IWS) is a term frequently used to define an unreliable piped urban water supply service that does not provide water to consumers every day of the week or 24 h a day (Charalambous & Laspidou 2017; Simukonda *et al.* 2018). Although IWS is considered the most accessible and cheapest short-term solution for meeting the increasing demand for water, poor governance and poor system management in the long term are the leading causes of perpetuating its intermittent nature (Charalambous 2012; Simukonda *et al.* 2018). More detailed information on the causes, consequences, and solutions of IWS can be found in the literature (Klingel 2012; Galaitsi *et al.* 2016; Simukonda *et al.* 2018).

IWS is commonly used in developing countries where water scarcity and other limiting factors are issues and/or will be an issue in the near future (Abu-Madi & Trifunovic 2013; Kumpel *et al.* 2017). Half of the water distribution systems in Asia, and two-thirds in Latin America, Africa, and the Middle East (Vairavamoorthy & Elango 2002; Vairavamoorthy *et al.* 2007; Klingel & Nestmann 2013) are operated intermittently. It has also been practised in some parts of Europe, mainly during dry seasons or periods of water scarcity (De Marchis *et al.* 2010). Around 1,313 million people are currently subjected to IWS, representing 18% of the world's population (7,301 million), 38% (2,774 million) have continuous water supply (CWS), and the rest (44%) lack a piped system altogether (Charalambous & Laspidou 2017). Different operational characteristics of IWS are applied in different areas; for instance, in 2009, the average supply time was 5.2 h in India and 8.5 h in Jordan (Danilenko *et al.* 2014). More detailed information can be found in Kumpel & Nelson (2016) and Charalambous & Laspidou (2017), in which the data for the operation of IWS systems around the world are synthesised. Galaitsi *et al.* (2016) specify the type of intermittency by proposing three definitions: (1) *Predictable* in which the supply schedule is known by consumers, and the amount of water received by each one of them is enough to cover their needs similarly to CWS systems. (2) *Irregular* where the supply schedule is unknown, but sufficient water can be collected by the user similar to the predictable type of intermittency. (3) *Unreliable* in which the water supply schedule is unknown and the amount of water mostly insufficient.

Intermittent supply results in water quality problems (Kumpel & Nelson 2016), pipe bursts, significant water losses (Mutikanga 2012; Erickson 2016), inequitable distribution (De Marchis *et al.* 2010; Chandapillai *et al.* 2012; Ameyaw *et al.* 2013), meter malfunctioning (Arregui *et al.* 2006; Criminisi *et al.* 2009; Mutikanga *et al.* 2011; Mutikanga 2012; Arregui *et al.* 2013; De Marchis *et al.* 2013; Fontanazza Chiara *et al.* 2015) along with coping cost for consumers, utilities, and the society (Vairavamoorthy *et al.* 2007; Ameyaw 2011). Negative consequences occur due to the unique characteristics of the system, such as supply interruptions which will be discussed later in this paper.

Understanding the system behaviour through hydraulic model simulations is crucial to proposing solutions for the aforementioned problems. In the absence of suitable software for IWS hydraulic modelling, researchers have explored different approaches. This paper aims to scrutinise the existing literature regarding hydraulic modelling and its application to IWS case studies and identify key gaps in knowledge about IWS systems. The implications of this paper are believed to pave the way for future studies in the context of IWS hydraulic modelling and, consequently, its improvement.

The paper consists of five sections: the second section, after the introduction, explains the approach adopted to do this review. The third section summarises and critically analyses the studies on hydraulic modelling methods of IWS developed and applied so far. It starts with clarifying the unique characteristics of IWS, followed by presenting the studies implemented in EPANET, mathematical models for simulation of the filling process and household tanks, macroscopic models, and methods applied in EPA-SWMM. The fourth section presents the issue of inequity in water distribution in IWS and how it has been addressed by different researchers. The final section presents the conclusions, key gaps in the body of knowledge, and recommendations for future research.

## 2. REVIEW METHODOLOGY

This review covers studies on IWS which take into account the hydraulic modelling aspect of such water supply systems. The first studies conducted on the topic date back to the late 1980s and early 1990s (e.g., Reddy & Elango 1989; Chandapillai 1991; Vairavamoorthy 1994), while interest and publications in the area of research have increased significantly lately. Web of Science, Google Scholar, SCOPUS, and ScienceDirect are the databases used. The journal papers retrieved from the databases were selected using the following search keywords: IWS, pressure-driven (dependent) analysis, hydraulic modelling, filling process, water distribution networks, water supply networks, and any combinations of these. There are a large number of publications presenting different pressure-driven analysis (PDA) methods with a wide range of purposes and applications (i.e., leakage, IWS, reliability analysis, design, and real-time control of water distribution networks (WDNs)). While many of these could be potentially used for modelling the pressurised state of IWS, only the ones that have already been used in IWS hydraulic modelling are referenced. Another key part of this paper refers to the filling process, which is a core element of IWS and a relatively new area of research interest.

The collected publications were analysed and discussed based on the following key aspects:

§ In terms of model building procedure:

  o The level of detail considered in the hydraulic model and how well it corresponds to reality.

  o The assumptions and simplifications in the methodology.

  o Efficiency in terms of computational cost and accuracy of the results (model errors) obtained in the context of the scale of the problem and network size.

  o Challenges faced in each case study and how they were addressed in hydraulic modelling.

  o Parameters and calibration, in order to acknowledge the unique characteristics of each case study and discuss their impact on results.

§ In terms of research objectives: the purpose of hydraulic model development and their applications, and the areas that the hydraulic models are developed and applied (e.g., improvement of supply equity, IWS to CWS conversion, leakage, etc.)

## 3. HYDRAULIC MODELLING OF IWS SYSTEMS

This section discusses the hydraulic modelling of IWS, which serves as a basis for the paper, beginning with a description of the characteristics of the IWS system itself.

### 3.1. IWS system characteristics

IWS systems need to be hydraulically analysed considering their unique nature, and this makes the modelling complex:
**Parameters for the hydraulic analysis of IWS** (Batish 2003):

1. *Water supply from the source*: Water is released from the source intermittently.
2. *Network charging process*: Since the water is distributed to the system at specified intervals, filling and emptying of the pipes occur periodically in the system (De Marchis *et al.* 2010). Dual-phase flow emerges (i.e., free-surface flow and pressurised flow) during filling/emptying.
3. *Household tanks*: During periods of disruption in water supply, some consumers store water in household tanks located either on their roofs or underground. It is common for consumers to store as much water as possible in their tanks when supply resumes while continuing to consume water for their everyday needs.

Contrary to CWS, in many IWS systems, pressures are often insufficient to satisfy consumers' demands at all or a number of demand nodes especially during peak demand hours. In such cases, outflows at demand nodes are dependent on pressure and the most appropriate approach is PDA. In PDA, pressure is considered explicitly in the system of hydraulic equations, and it is directly linked with demand. PDA methods achieve this by incorporating a nodal head-flow relationship (NHFR) in the system of hydraulic equations. Several NHFRs have been proposed in the literature with the distinguishing factor being the existence of an upper boundary for outflow in their equations. Equations with an upper boundary (Bhave 1981; Germanopoulos 1985; Wagner *et al.* 1988; Chandapillai 1991; Fujiwara & Ganesharajah 1993; Tanyimboh Tiku *et al.* 2001; Tanyimboh & Templeman 2010) are suitable for modelling demands with controlled outlets (e.g., taps or tanks with flow control mechanisms or float valves). On the other hand, equations without an upper boundary (Reddy & Elango 1989, orifice equation) are suitable for modelling demands without a control mechanism (e.g., uncontrolled outlets or overflowing tanks).

Upper and lower boundaries ($H_{\min}$, $H_{\mathrm{req}}$) in NHFRs designate the pressure head above which pressure-dependent outflow occurs and the pressure head above which full demand is supplied, respectively. Consequently, actual consumption at demand nodes (taps or tanks) differs from the original demand. The actual flow delivered at each demand node (DN) is based on the amount of water released from the source, the pressure availability, and user behaviour (Batish 2003; Ingeduld *et al.* 2006; De Marchis *et al.* 2010). Thus, under such conditions, it is difficult to predict the user water consumption.

Shirzad *et al.* (2013) investigated various pressure–discharge relationships with experimental work. Different size faucets with different opening status were explored. Results showed that the orifice equation and Wagner *et al.*'s (1988) equation without the upper threshold for which full demand is delivered showed better agreement with the experimental data. Tanyimboh & Templeman's (2010) equation showed good agreement with collected results after the calibration of parameters $\alpha$ and $\beta$. Walski *et al.* (2017) performed an experimental study in a laboratory set-up trying various pipe layouts and orifice openings. Results showed that the observed pressure and flow outputs agreed well with the orifice equation model. In further experimental work of Walski *et al.* (2018), the impact of having outlets at different elevations at a DN was investigated. Results showed that the elevation differences of outlets can have a significant impact on outflows. The proposed equation that describes the exponent takes into account the number of orifices at the DN, the range of orifice elevations, and the elevation below the node of the lowest orifice.

Experimental studies in laboratories could contribute to improved understanding and give insights into system behaviour under pressure-deficient conditions and complex consumer behaviour. A major advantage of laboratory studies is the flexibility and simplicity of arrangements as well as the ease of simulating different scenarios in a time-efficient manner compared to real case studies. Full control of the system and reasonably accurate data acquisition are important characteristics that could lead to the validation of current concepts as well as new findings. An obvious disadvantage of laboratory case studies is their small scale, although with current knowledge being limited modelling of more simple cases seems reasonable.

It is a common practice in CWS models that demand points and their corresponding consumption are aggregated at the model's DNs. Major assumptions about elevations and connectivity to the main WDN are made. Such simplifications could be reasonable in the case of sufficient pressure that guarantees CWS. However, the results of the above NHFR experimental studies suggest that in a range of low pressures, elevation differences in outlets can significantly affect outflow (Walski *et al.* 2018). Also, other parameters that increase the complexity of the relationship between flow at a node and pressure are the variable human behaviour of the person opening the faucet, the difference in elevations of the faucet and the model node, and the complexity of piping between them.

**Challenges for building an IWS model:** Data acquisition is one of the most significant challenges in building the model. Building a decent quality model depends on representing all elements of the system as realistically as possible. While in most CWS cases, design and maintenance interventions are well documented, in some IWS cases, the position, connectivity, and status of hydraulic elements are unknown and need to be mapped before building the model. Finally, model building is exacerbated by the lack of publicly available software suitable for IWS simulation.

**Challenges for calibration and validation of the methods:** The availability of measurement data is a fundamental issue in the analysis of IWS as it is imperative for calibration and validation as well as model building. After building an accurate model of the WDN, it is equally essential to calibrate and validate its correspondence to the real system based on pressure and flow-field measurements.

**Issues resulting from the operational management of IWSs:**

§ In the event of repeated opening and closing of valves, there can be pressure fluctuations leading to pipe failures. Other than pressure fluctuations, dynamic phenomena like transient flows are generated contributing in pipe and fittings' wear. Consequently, an increase in leakage occurs in such systems (Klingel 2012; Dighade *et al.* 2014; Charalambous & Laspidou 2017).

§ There is an increase in apparent losses caused by meters under registration in two ways: first, meter aging, and second, due to the low-flow rate induced by float valves in households (Criminisi *et al.* 2009; Mutikanga *et al.* 2011).

§ Illegal connections are triggered by a lack of water availability, poor management by the utility, breakdown of recording consumer registration, corruption and mismanagement, lack of household connections, and inappropriate tariffs (Butler & Memon 2006; Vermersch *et al.* 2016). It is estimated that 50% of water losses are due to private and illegal connections (Galaitsi *et al.* 2016).

Vairavamoorthy (1994) categorised IWS systems based on water availability as starved and non-starved systems. 'Starved' defines systems where supplied water is not enough for consumers to fill their household tank. The term 'non-starved' describes systems where sufficient water can be collected to fill household tanks during supply duration. Similarly, Taylor *et al.* (2019) defined IWS as satisfied and unsatisfied based on the amount of water a single consumer can get. If a consumer can get as much water as they demand, they are called 'satisfied'; if not, they are called 'unsatisfied.' A network can also consist of satisfied and unsatisfied users. These classifications are crucial for the analysis and assessment of IWS. Both classifications have similar meanings and are appropriate for representing IWS.

To date, there is no available software for the simulation of the hydraulic behaviour of IWS. In the absence of suitable software for modelling, researchers have proposed different methods to analyse the system. Most studies focus on the pressure dependency of nodal outflow and use different PDA methods mainly through EPANET (Section 3.2). However, some researchers use EPA-SWMM that was originally developed to model free-surface flow in drainage systems. The use of EPA-SWMM for IWS modelling involves manipulating the available hydraulic elements (nodes, conduits, etc.) to behave according to hydraulic elements in a clean WDN (Section 3.4). Other researchers developed new mathematical models that include the filling process (Lieb 2016) and the existence of household tanks (Criminisi *et al.* 2009) (Section 3.3.). An overview of IWS hydraulic modelling approaches is shown in Figure 1.

## 3.2. Modelling IWS in pressurised state (EPANET)

This section presents and discusses studies that have proposed and applied various pressure-dependent demand (PDD) hydraulic models that implement NHFRs in real case studies. All the following studies focus on modelling the fully pressurised state of the WDN while the filling and emptying processes are not considered. The main focus is to include intrinsic elements of an IWS system, such as low pressures at DNs and the consumers' household storage tanks. In most cases, the aforementioned PDA methods (Section 3.1.) are used to take into account the effect of low pressures on the actual demand outflows.

### 3.2.1. Controlled outlets

Trifunović & Abu-Madi (1999) tested and compared two different approaches for modelling demands in IWS systems where consumers use household storage tanks. The first approach uses a pressure threshold at which the specified demand is satisfied and below which the outflow is pressure-dependent. Different pressure thresholds were tested (6, 12, and 36 m) in a case study, and results showed instability in the pressure profiles at DNs as well as negative pressures. In the second approach, the individual household roof storage tanks were clustered and represented by one large but shallow tank (depth of 1–2 m) in the network model. A check valve (CV) and a flow control valve (FCV) were also added between the original DN and the artificial tank to prevent backflow to the system and limit outflow at DNs to the required demand, respectively. The second approach that includes the introduction of tanks is more realistic than the first approach, which produces profound changes in pressure and even negative pressures.

Similar to the work of Trifunović & Abu-Madi (1999), Macke & Batterman (2001), and Ameyaw (2011), individual household storage tanks were aggregated into larger-sized tanks in the model. In Macke & Batterman (2001), the size of the

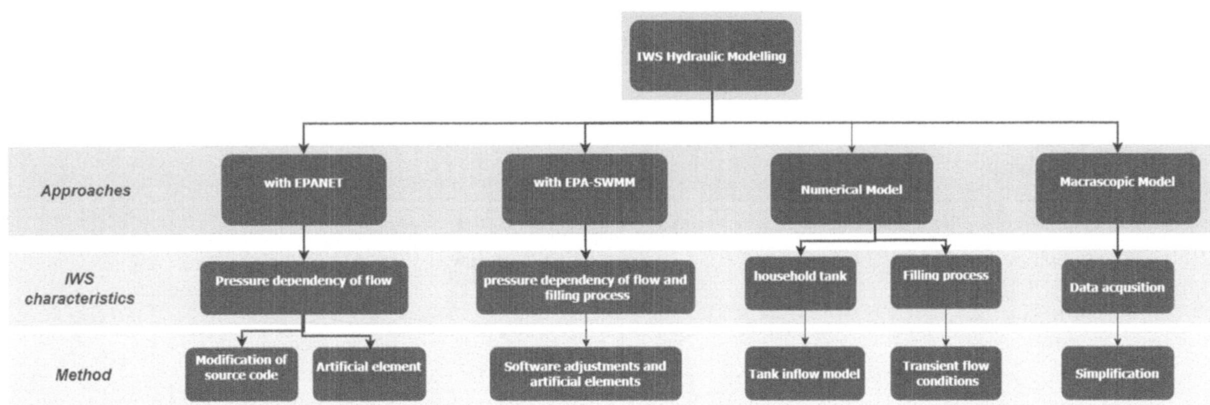


**Figure 1** | IWS hydraulic analysis methods.

equivalent model tanks and the diameter of the pipes connecting the tanks to the network were subject to calibration using field data. The equivalent tank depth is 1 m for all tanks, and a CV is introduced between the original DN and the tank. A set of artificial elements was also used in order to model pressure-dependent leakage. A 2-m long pipe leading to a tank of a very big size is connected to all leakage nodes. Calibration of the model was conducted on multiple factors, including the diameter of artificial leakage pipes, instead of the usual calibration, which considers only pipe roughness.

Ingeduld *et al.* (2006) incorporated Wagner *et al.*'s (1988) NHFR and considered household storage roof tanks in their hydraulic model. The assignment of tanks to network junctions is done in a semi-automated way using available geographic information system (GIS) data and consumer questionnaires (e.g., for the physical properties of the tanks and their connection to the network). Leakage is taken into account as a percentage of the total supplied volume. In Puleo (2014), household tanks' fill rate is modelled with Wagner *et al.*'s (1988) equation and depends on the level of the tank's floater level. $H_{min}$ is considered equal to the tank elevation and $H_{req}$ is assumed to be equal to $+10$ m after considering the head losses in the consumer connection pipe.

A summary of the studies mentioned in this section can be found in Table 1.

**Table 1** | IWS studies using EPANET for hydraulic modelling considering controlled outlets

| | Hydraulic model | | | | | |
| Authors | PDD model | Leakage model | Private tanks | Study objective | Details of objective | Case study |
|---|---|---|---|---|---|---|
| Trifunović & Abu-Madi (1999) | Tanks or NHFR | – | – | Demand modelling of the IWS system with private storage tanks | Individual household RSVs are modelled either as PDD nodes, or as shallow tanks of large surface area | Tulkarm, Palestine |
| Macke & Batterman (2001) | Tanks | Artificial pipe tank arrangement | | Calibration, simulation, and water loss reduction in IWS | Individual household tanks are aggregated at nodes and are represented by large tanks in which size and connecting pipe diameter are calibrated with field data | Al Koura, Jordan |
| Ingeduld *et al.* (2006) | Wagner *et al.* (1988) | Distributed across network pipes as % of demand | Tanks | PDA of IWS | Provide a PDA tool to help in network rehabilitation, planning, and conversion to CWS | Shillong, India and Dhaka, Bangladesh |
| Ameyaw (2011) | Tanks | – | – | Macroscopic hydraulic simulation of IWS with the use of tanks and equity improvement interventions during design | Installation of system tanks is considered in order to improve the equity of water distribution. Optimisation of tank size based on the maximisation of equity and the minimisation of cost | Four nodes in series network (Gupta & Bhave 1996) |
| Puleo (2014) | Wagner *et al.* (1988) combined with orifice equation | – | – | Real-time optimal control of IWS | Novel hydraulic model for IWS with household tanks and optimal valve operation | Palermo, Italy |
| Ilaya-Ayza *et al.* (2017) | Emitter (modified EPANET2 Pathirana (2010)) | – | – | Multi-criteria optimisation of supply schedules in IWS | Criteria for optimisation are to: maintain pressures across the network close to a reference pressure, reduce inconvenience for consumers, modify the shorter supply schedules, and ease sector operation | Oruro, Bolivia |

### 3.2.2. Uncontrolled outlets

Batish (2003) and Chandapillai *et al.* (2012) investigated the design of IWS systems aiming to provide equitable distribution of water resources to consumers. Batish's (2003) demand model is based on attaching at each DN a reservoir (RSV) with an elevation of the required total head. The author notes that modelling demands with tanks would be more realistic (as in Trifunović & Abu-Madi (1999) and Macke & Batterman (2001)). However, this approach would require much more detailed information about the individual household storage tank sizes and their connection to the network. Chandapillai *et al.* (2012) consider pressure dependency of the delivered demand through the NHFR proposed by Reddy & Elango (1989) and Chandapillai (1991). This was incorporated into the EPANET model by adding an emitter at each DN, which substitutes the assigned demand.

Fontanazza *et al.* (2007) considered that consumers use household storage tanks with pumps to collect as much water as possible during the supply period. The pump ensures the outflow of water in the storage tank even when the nodal pressure is less than the minimum required ($H_{\min}$) for having an outflow at the node. Reddy & Elango's (1989) NHFR is used to model the pressure dependency of outflow in household tanks. Due to the presence of the pump, there is no minimum required pressure to have outflow at the storage tanks ($H_{\min} = 0$). It is also considered that there is no $H_{req}$ for which the demand is satisfied. The total nodal outflow is limited only by the tank size which is defined according to the nodal daily demand. The pump characteristics are chosen, so that it can fill the storage tank in 4 or 5 h.

In Mohapatra *et al.* (2014), artificial RSVs are assigned at each DN with an elevation equal to the ground level plus the required pressure head. Outflow to the artificial RSV in this case is not restricted by setting an upper boundary of required pressure that satisfies the full demand. Each artificial RSV in the model is connected to the main network with a pipe of diameter and length equivalent to that of the pipes leading to consumer ends. The length and roughness coefficient for the equivalent pipe is that of the dominant pipe and its diameter is determined appropriately as in Walski *et al.* (2003). This pipe is also set to act as a CV to prevent backflow from the RSV to the main distribution network. Leakage is also taken into account in the model as discharge through emitters with a coefficient of 0.6. The emitter coefficient is calibrated with leakage data that assume equal distribution of the total leakage flow rate across leakage points.

Gottipati & Nanduri (2014) used Tanyimboh Tiku *et al.*'s (2001) NHFR but without an upper threshold for outflow. The methodology for equity assessment and the hydraulic model were applied in the real case study used in Kansal *et al.* (1995), as well as two sample networks, one with a grid layout and one with a radial layout. In Ilaya-Ayza *et al.*'s (2017) supply schedule optimisation study, hydraulic simulations are performed using the EPANET 2 version as modified by Pathirana (2010). High peak flows cause reduction of pressure and flow at the most remote or elevated points relative to the source.

Table 2 summarises the studies mentioned above. These studies use a wide variety of hydraulic modelling techniques depending on each specific case study. In cases where consumers were considered to leave their taps open throughout the supply period (uncontrolled outlets), no specific demand was attributed at DNs. Uncontrolled outlets are taken into account by using either PDD equations without head pressure upper limit for which demand is fully satisfied ($H_{req}$) or artificial RSVs at DNs which maintain a constant elevation and do not fill. In other cases, consumers' behaviour is considered by specifying nodal demands or assigning finite volume tanks at DNs. Modelling nodal demands with tanks is realistic in cases where consumers use storage tanks. This can be considered a volume-dependent demand, and the tank's filling rate is pressure-dependent. In cases where consumers use tanks that overflow, the approach of no upper limit in outflow would be more realistic.

## 3.3. Numerical, empirical, and macroscopic models

Some researchers have developed complex mathematical models considering the unsteady filling process and the existence of the household tanks to represent the hydraulic behaviour of the system more accurately. Others have proposed more simplistic approaches (i.e., macroscopic models) while addressing the challenge of getting detailed data from real IWS systems. These are presented under the following subsections.

### 3.3.1. Integration of periodic filling and emptying process in the hydraulic analysis of IWS

With the intermittency of water supply, periodic network filling and emptying of the piped network occurs in IWS, which then leads to the development of transient flow phenomena with highly varying velocity and head (i.e., from pressurised to partial-flow regime and *vice versa*). Such behaviour points to the necessity of integration of continuity and momentum equations (Ingeduld *et al.* 2006).

**Table 2** | IWS studies using EPANET and considering uncontrolled outlets for hydraulic modelling

| Authors | Hydraulic model | | | | | |
| | PDD model | Leakage model | Private tanks | Study objective | Details of objective | Case study |
| --- | --- | --- | --- | --- | --- | --- |
| Batish (2003) | RSVs | – | – | Design of IWS systems | RSVs are used to model demands. Different positions for FCV installation are investigated in order to improve the equity of water distribution | Pinjore, Haryana, India |
| Fontanazza *et al.* (2007) | Reddy & Elango (1989) (emitter) | – | Pump and RSV | Analysis of IWS and assessment of water distribution equity | Investigate the equity of water distribution in the case where consumers use private pumps to store water in household tanks | Palermo, Italy |
| Chandapillai *et al.* (2012) | Reddy & Elango (1989) (emitter) | Distributed across network pipes as % of demand | Tanks | Design of IWS systems for the equitable distribution of water | Optimisation of the different pipe diameters based on the maximisation of equity and the minimisation of cost | Two-loop sample network |
| Mohapatra *et al.* (2014) | RSVs | Emitter | – | Network assessment in IWS and CWS | Model IWS with uncontrolled outlets at consumers' end with RSVs at DNs | Nagpur, India |
| Gottipati & Nanduri (2014) | Wagner *et al.* (1988) | – | – | Assessment of equity in water distribution and optimal design of IWS systems | Assess the impact of source elevation, variation of demands at DNs, and changes in pipe diameters in equity | Kansal *et al.*'s (1995) case study |

IWS systems may be more accurately modelled if the network charging and emptying processes are also included in the analysis. By doing so, the moment each consumption node starts receiving water will be taken into account. This crucially impacts the unfair distribution of water among consumers.

Water network charging (i.e., hydraulic filling of pipes) has been modelled either using the Rigid Column Theory (RCT) or Elastic Column Theory (ECT) (Walski *et al.* 2003; Bhave & Gupta 2006; Zhou *et al.* 2011). While in RCT, it is assumed that the water is incompressible and the pipes are rigid; in ECT, the elasticity of the pipes and the compressibility of the fluid are considered as it assumed to be the result of momentum variations (Walski *et al.* 2003; Bhave & Gupta 2006).

RCT has limited applications in analysing instantaneous flow changes as it is unable to interpret the behaviour of pressure wave generation. Transient flows are only controlled by inertia and the friction that is applicable to surge conditions (i.e., in the case of slow and minor head changes, the compressibility of the flow and pipe deformation can be neglected) (Walski *et al.* 2003; Bhave & Gupta 2006). Considerations of pipe elasticity and flow compressibility in ECT produce wave propagation characterised by the elasticity of the liquid and the pipeline (Walski *et al.* 2003; Bhave & Gupta 2006).

The boundaries of the system (e.g., tanks and dead ends) play a crucial role as well as equations that describe the transient flow, as they also impact the behaviour of the transient conditions (Walski *et al.* 2003).

To consider the unsteady filling process, Liou & Hunt (1996) developed a numerical model which is based on the RCT. The model was verified with a laboratory test. However, De Marchis *et al.* (2010) claim that the model is not suitable for complex WDNs due to the fluid incompressibility assumption. They proposed a model using the method of characteristics (MOC) to solve the partial differential equations of momentum and continuity for weakly compressible liquids and applied to the city of Palermo to calibrate and validate it. They combined the tank inflow emitter law of Criminisi *et al.* (2009) and the filling process equation based on the assumption that the water column cannot be fragmented and the air pressure internal to the network is equivalent to the atmospheric pressure and created a mathematical model. The average node head discharge parameter settings are inspired by Criminisi *et al.*'s (2009) field study. Their results seem promising in terms of evaluating the strategies for intermittent WDN management, particularly addressing the inequitable supply problem. De Marchis *et al.* (2011) then used this numerical model to analyse the inequitable distribution of water under different water availability situations. However, from Mohan & Abhijith's (2020) point of view, this model has limited application to small diameter pipes because they assume that the water column front is perpendicular to the pipe axis and the flow depth is always equal to the pipe diameter.

Additionally, their model is not made available, and its validation has been done with data of a single time step. Freni *et al.* (2014) used this hydraulic model in their proposal of using pressure reduction valves (PRVs) for improving water supply equity. Other numerical models employing the transient flow conditions are developed by Nyende-Byakika *et al.* (2012) and Nyende-Byakika *et al.* (2013).

Lieb (2016) developed a mathematical model to capture the bi-dimensional behaviour of the filling's front as a mathematician using the Preissman slot model. This model ensures a smooth transition between the empty and the pressurised pipes, and at the same time, it incorporates time-dependent boundary conditions. To reduce the risk associated with infrastructure damage, optimisation that improves the system operations is proposed. However, the available computer simulations are computationally expensive due to the complexity of such models.

Experimental works were also carried out (Walter *et al.* 2017; Weston *et al.* 2022) using one pipe connected to an RSV. Walter *et al.* (2017) investigated the metering errors caused by air resulting from pipeline filling. Weston *et al.* (2022) explored the effects of some hydraulic parameters (initial/final pressure, initial/final flow rate, valve operation speed, and leakage percentage) on the filling and emptying of pipes. They observed the negative pressures and over-pressurisation of the system, which can be associated with the water quality. Walter *et al.* (2017) observed that 93% of the air volume in the pipe is measured by the meters due to the filling process of an empty service connection.

Mohan & Abhijith (2020) proposed an improved PDA hydraulic model that considers partial-flow regimes inside the piping system of the network. Integration of partial flow in the hydraulic model is a novel feature in IWS hydraulic modelling. Partial flows in pipes are common either during the filling process or at later stages of supply. Their study focuses on highly intermittent WDNs, and thus, the head–outflow relationship considers outflows through uncontrolled outlets as in Reddy & Elango (1989). The model was tested on the real case study of the WDN of North zone 10 of Madurai, Tamil Nadu, India, which consists of 30 DNs and 30 pipes in a branched configuration. The model is useful for both pressurised and pressure-deficient conditions, and the outcomes can be fruitful for the analysis of water quality variations (Mohan & Abhijith 2020).

The studies mentioned in this section are shown in a summary form in Table 3.

The steady-state analysis is sufficient to analyse hydraulic systems that undergo slow changes in velocity and pressure. A transient analysis is required when changes in flow and pressure occur rapidly within a short period of time. Transient flow conditions were developed in IWS during the network filling process. Rapid filling problems are two-phase phenomena, and it is necessary to implement multi-phase modelling (Vasconcelos & Wright 2005). However, multi-phase modelling is more complex, and the computed results would have more uncertainty. An accurate theoretical simulation of every physical phenomena that can occur in an actual hydraulic system is nearly impossible. Therefore, all modelling of transients involves some approximation and simplification of the real problem (Walski *et al.* 2003).

The transient phenomenon should be considered not only in the analysis process but also in the design process. The operating speed of the valves and the size of the pipes should be chosen, so that the detrimental effects of transients can be avoided.

### 3.3.2. Macroscopic models

Taylor *et al.* (2019) approached the modelling aspects of IWS from a unique perspective proposing a parsimonious model addressing the difficulty in detailed data acquisition from such networks. The model aims at understanding the operation and persistence of IWS by indicating the relationships among consumer demand satisfaction, source water availability, consumer demand, and leakage. Aggregated behaviour of a system is considered in this macroscopic model. Hence, the entire system is represented based on user behaviour, and a single leakage with a mathematical equation based on the volume of water supplied to the network was taken as equal to the sum of the volume received by consumers and the leakage volume:

A single satisfied and unsatisfied consumer: $\quad V_R = \begin{cases} V_D & \text{:Satisfied} \\ K_D\, th^{\emptyset} & \text{:Unsatisfied} \end{cases}$ $\hfill$ (1)

Aggregated consumer: $\quad V_R = V_D \min\left(1, \dfrac{th^{\emptyset}}{\gamma_S}\right)$ $\hfill$ (2)

**Table 3** | Numerical and empirical studies for filling and emptying process of IWS

| Authors | Category | Used model | | | Study objective | Methodology | Application/case study | Conclusion |
|---|---|---|---|---|---|---|---|---|
| | | Theory and solution method | Filling process | | | | | |
| | | | 1D | 2D | | | | |
| Liou & Hunt (1996) | Unsteady filling process | RCT | | – | To consider the unsteady filling process | A numerical model was developed, and it was verified with a laboratory experiment | A pipeline connected to a RSV | The proposed model can be applied to the networks; in case, there is no air intrusion: high-flow velocity or small diameter of pipes |
| De Marchis et al. (2010) | IWS filling process | Both water and pipeline characteristics are considered by using the MOC | | – | To simulate the IWS filling process where users have household tanks | A mathematical model was developed for the simulation of the filling process that takes place in IWS | One of the supply networks in Palermo/Italy | The occurrence of inequality due to household tanks was confirmed with the model applied to a real case study. The developed model can be useful for further analysis of IWS |
| Nyende-Byakika et al. (2012) | Pipe filling process | Different solution methods are applied for different flow conditions | | – | To perceive flow transition from free-surface flow to pressurise flow takes place filling of IWS | Equations derived from Saint Venant's principle considering spatial and temporal changes in flow and pressure characteristics | – | The proposed algorithm helps understanding of low-flow conditions by calculating the pressure and flow at a particular time |
| Nyende-Byakika et al. (2013) | IWS system analysis | MOC | | – | To develop models for IWS, particularly free-surface flow conditions, namely low pressure open channel flow (LPOCF) | Three approaches were suggested addressing IWS: demand-dependent analysis (DDA), PDA and modelling the pipe filling and pressurisation | WDN of Kampala/Uganda | The framework suggested for IWS will be useful for operational considerations |
| Freni et al. (2014) | IWS filling process | MOC | | – | To develop a dynamic model that accounts for the filling process and the existence of PRVs | The model of De Marchis et al. (2011) (network filling process) and the dynamic approach of Prescott & Ulanicki (2008) (PRV model) were combined | One of the WDS of Palermo in Sicily | The importance of PRV usage to improve equity was demonstrated |
| Lieb (2016) | IWS filling process | Preissman slot model | | – | To capture the two-dimensional behaviour of transient flow during the network filling process | A numerical model was developed to enable the transition from empty to pressurised pipes, and then optimisation was used to enhance IWS | Alameda/Arraiĵ́an/Panama | The work could be the starting point for building modelling and optimisation tools for IWS |

| | | | | | | | | |
|---|---|---|---|---|---|---|---|---|
| Walter *et al.* (2017) | Experimental pipe filling process | Experiment | – | – | To measure the error in a single jet water metre caused by air during the filling process | Experiments were conducted for dry and wet metre case | A pipe connected to a water tank | While measurement error only depends on air volume in the dry metre case, additionally it depends on pipe pressure in the wet case |
| Weston *et al.* (2022) | Experimental pipe filling process | Experiment | – | – | To perceive the effects of some hydraulic parameters on the duration of the filling and emptying process | Inequitable distribution was tested under different scenarios | A single pipe connected to a RSV | Negative pressures were observed during the filling process |
| Mohan & Abhijith (2020) | Partial-flow regimes in IWS | PDA-PF model developed | | | To introduce a methodology accounting for partial-flow regimes in IWS | The PDA-PF model was proposed. Starved network conditions were elucidated by including uncontrolled outlets in the model. Simulation results from EPANET were compared with PDA analysis | Madurai city, India | The duration of the partial-flow conditions was found to make up a quarter of the total duration |

where $V_R$ (m³/day) and $V_D$ (m³/day) are the daily volume of water received and demanded by consumers, respectively. $K_D$ (m³/day) is pipe and topography constant, $t$ is duty cycle (fraction of time a system is pressurised), $\varphi$ is the pressure exponent of consumer demand, $h \equiv H/H_t$, $(H)$ supply pressure, $(H_t)$ targeted supply pressure, and $\gamma_S$ is the minimum $th^\varphi$ required to satisfy consumers.

Leakage rate of a single leak:   $Q_L = C_d A [2gH]^\alpha \left( \dfrac{86,400\ S}{1\ \text{day}} \right)$ $\qquad\qquad\qquad$ (3)

Daily leakage volume for the network:   $V_L = tQ_L = K_L A t H^\alpha$ $\qquad\qquad\qquad$ (4)

Model equation:   $V_P = V_D \min\left( 1,\ \dfrac{th^\varnothing}{\gamma_S} \right) + V_{LC}\alpha th^\alpha$ $\qquad\qquad\qquad$ (5)

$Q_L$ (m³/day) is the leakage flow rate of a single leak, $A$ is the orifice's cross-sectional area, $C_d$ accounts for the orifice's shape, $\alpha$ accounts for the flow rate's pressure dependency, $V_L$ (m³/day) and $V_{LC}$ (m³/day) is the daily volume of leaked water and targeted leaked water, $V_P$ (m³/day) is the daily volume input to a network, and $\alpha \equiv A/A_t$, $A_t$ is the equivalent orifice area (EOA) required to achieve a project or system's targets.

The model is calibrated using four reference CWS networks that are simulated as IWS using the method of Macke & Batterman (2001). The model indicates that there is an optimum point between satisfied demand and unsatisfied demand which might be the justification for the existence of IWS. On one hand, the model is extremely useful for the improvement of IWS, such as conversion purpose (from IWS to CWS), and it can be applied to different IWSs. On the other hand, this macroscopic model has some limitations to represent IWS systems. Lack of network topology and spatial variation of pressure affect other system parameters dependent on pressure such as the uniqueness of individual consumers, the gradual transition from unsatisfied to satisfied, or the pipe filling process (Taylor *et al.* 2019).

Similarly, in 2018, a simplified model was proposed by Taylor *et al.* (2018) to explore strategies for the improvement of water quality in IWS. The model equations relate supply duration, supply pressure, and leakage rate. They represented leakage with Equation (4), in which the EOA was used to quantify water quality. They concluded that longer supply duration might improve water quality but only during the flushing phase. In the subsequent steady state, it creates an adverse effect.

### 3.4. Modelling of IWS with EPA-SWMM

EPA-SWMM has been preferred for hydraulic analysis of IWS by some researchers, as it is capable of simulating the transition from the free surface to pressurised flow.

The usage of EPA-SWMM started with a masters thesis written by Segura (2006) targeting the conversion of IWS to CWS. The idea of developing his method in EPA-SWMM was to represent the pressure dependency of flow related to the existence of household tanks. For this reason, a storage tank is connected to the node with an outlet at each junction, in which the pressure dependency of flow is assigned. The method also includes outflow from storage with a normal daily water consumption curve assigned as negative inflow to the storage nodes. The application of this method to the case study in Villavicencio in Colombia showed that the inequitable distribution is exacerbated by the usage of bigger tank size.

Similarly, Cabrera-Bejar & Tzatchkov (2009) included household tanks in their method but simulated supply schedules of IWS by introducing an inflow pattern to the source node. Kabaasha (2012), Shrestha & Buchberger (2012), and Dubasik (2017) developed methods in EPA-SWMM to simulate IWS behaviour with slightly different node configurations and adjustments of settings in the software. Whereas Kabaasha (2012) and Dubasik (2017) preferred to dismiss the inclusion of household tanks in node configuration by assuming that users always leave their taps open, Shrestha & Buchberger (2012) included a pump in the node configuration as well as household tanks. That way, the inflow at each node is dependent on the available nodal pressure and the pump capacity. Kabaasha (2012) implemented his assumption to the node configuration by connecting an outfall to each DN via an outlet. Thus, the inflow will only depend on the available pressure at the node, and it is implemented by introducing an NHFR at each outlet.

The ability of EPA-SWMM in representing the behaviour of the filling process was evaluated by validating the results with a field experiment in a WDN in Sicily (Italy) by Campisano *et al.* (2019a). Their method is the same as the method developed by Kabaasha's (2012) PDA approach, except that they introduced different head-flow relationship equations to the outlet.

Meanwhile Kabaasha (2012) used an exponential relationship, and Campisano *et al.* (2019a) used Wagner's equation (Wagner *et al.* 1988). They concluded that EPA-SWMM is reliable in simulating the filling process of IWS.

With the aim of considering household tanks in the modelling, Campisano *et al.* (2019b) developed a method including storage elements connected to the nodes. They used the hyperbolic law of the emitter developed by De Marchis *et al.* (2015) to predict the flow rate based on available pressure entering household tanks controlled by float valves:

$$\begin{cases} Q = Q_{max} = C_v a_v \sqrt{2g[(h - z_t) - SL]} & \text{if} \quad H \leq H_{min} \\ Q = Q_{max} \tanh\left(m\dfrac{H_{max} - H}{H_{max} - H_{min}}\right) \tanh\left(n\dfrac{H_{max} - H}{H_{max} - H_{min}}\right) & \text{if} \quad H_{min} < H < H_{max} \\ Q = 0 & \text{if} \quad H \geq H_{max} \end{cases} \tag{6}$$

where $Q$ (m$^3$/s) is inflow to tank; $Q_{max}$ (m$^3$/s) is the value of inflow when the float valve is fully open; $C_v$ is the float valve emitter coefficient; $a_v$ (m$^2$) is the valve inflow area; $g$ (m/s$^2$) is the gravity acceleration; $h$ (m) is nodal pressure head in the network; $z_t$ (m) is float valve elevation; $S$ is the energy line slope; $L$ (m) is the length of pipe connection between the network node and tank; $m$ and $n$ are non-dimensional coefficients; $H$ (m) is tank water level; and $H_{max}$ (m) and $H_{min}$ (m) are, respectively, tank water levels corresponding to full closure and full opening of the float valve.

This method (Campisano *et al.* 2019b) was employed for the improvement of equity by Gullotta *et al.* (2021).

Since EPA-SWMM requires at least an outfall to drain water from the system (Rossman 2017), methods do not have outfalls connected to the junction nodes placed an outfall somewhere in the network. Table 4 summarises the methods applied in EPA-SWMM for modelling IWS. All the methods use the dynamic wave routing method, which allows the simulation of pressurised flow by solving the St Venant equations (Rossman 2017).

Overall, the hydraulic analysis of IWS with EPA-SWMM served different purposes: simulation of the network filling process (Cabrera-Bejar & Tzatchkov 2009; Campisano *et al.* 2019a); more accurate consideration of household tank (Campisano *et al.* 2019b); hydraulic modelling of IWS with EPA-SWMM with the comparison of the EPANET-PDA approach (Kabaasha 2012); better management of coping strategies of users by having information on the time of getting water (Segura 2006); improvement of equitable supply (Dubasik 2017); and the provision of a reliable continuous supply (Shrestha & Buchberger 2012).

Methodologies were developed for specific case studies, and hence, settings are made accordingly. Each of the methods mentioned above approximates the IWS behaviour and is useful in terms of proposing solutions caused by intermittencies such as inequitable supply and the convergence of the system to CWS. However, they have challenges with detailed data acquisition. Furthermore, since EPA-SWMM requires substantial modification incapable of simulating the water distribution system (WDS), and the complexity of the underlying equations pose some errors in the model results (e.g., the existence of flooding the junctions, the appearance of a flow routing error, and increase in run time caused by embedded additional elements). The suitability of the settings in terms of the working principles of EPA-SWMM is crucial for both the accuracy and the stability of the results. Furthermore, most of the studies did not validate the results against the field data as data acquisition is one of the main challenges faced.

### 3.5. Observations

The different hydraulic models for IWS systems presented in this section can be addressed based on the level of detail they are emulating. Macroscopic models attempt to provide a system representation on the highest scale by aggregating the major components of water input and output such as volumes of water supply from source(s), demands, and leakage. Furthermore, more detailed models (e.g., EPANET) represent hydraulic elements in greater detail (e.g., specific demand points and household tanks) while taking into account pressure dependency of demands and household tank filling. Some models (e.g., EPA-SWMM and numerical models) are even able to simulate processes such as the filling process or partial flow in pipes.

Each of the above models can be useful for decision-makers for different purposes and uses. Their suitability can depend on data availability, uncertainty of the physical and operational status of elements in the real WDN being modelled (pipe diameters, valve statuses, leakage levels and location, etc.), and the desired level of accuracy or scale. It should be stressed that IWS shows increased complexity compared to CWS and poses different challenges in the modelling process. Current studies are still investigating different methodologies for modelling and analysis of such systems.

**Table 4** | IWS studies using EPA-SWMM for hydraulic modelling

| | Hydraulic model | | | | | | | |
| Authors | Artificial elements | NHFR | Private tanks | Filling process | Study objective | Details of objective | Case study | Supply schedule |
| --- | --- | --- | --- | --- | --- | --- | --- | --- |
| Segura (2006) | Outlets and tanks | Applied based on field measurements | Tank | 1D | Conversion from IWS to CWS | To learn the time water delivered to each consumer, so that they can plan the water usage. In this way, large storage usage can be prevented | Villavicencio, Colombia | 7 days |
| Cabrera-Bejar & Tzatchkov (2009) | Tanks | – | Tank | 1D | Modelling initial network charging | A schedule of IWS is introduced to the network | Guadalajara/ Mexico | 3 days – the first 5 h a day |
| Kabaasha (2012) | Outlets and outfalls | Linear relationship | – | 1D | Modelling unsteady-flow regimes in WDS | Various networks are used for both demand-driven and pressure-driven analysis; results are compared with EPANET 2.0 | Real-life and synthetic networks | 24-h simulation with the first 3 h and the last 6 h supply for unsteady-flow simulation |
| Shrestha & Buchberger (2012) | Pump tank outlet outfall | – | Tank | 1D | IWS conversion to continuous with satellite water tanks | Modelling of existing IWS and proposed WDS with satellite tanks | Kathmandu/ Nepal | Duration of simulation: 14 days water supplied every 2.5 h for alternate days in a week. The intermittent inflow of 2.6 litres per second (lps) is introduced at the source node |
| Dubasik (2017) | – | – | – | 1D | The transition from IWS to CWS and the improvement of equitable supply | A real case study designed for CWS is used. CWS operation is modelled with EPANET 2.0 with DDA. IWS operation is modelled with EPA-SWMM | Espavé/Panamá | Some scenarios are applied |
| Campisano et al. (2019a) | Outlets and outfalls | Wagner et al. (1988) | – | 1D | The ability of EPA-SWMM for simulating the filling process | The filling process validated with field data | Ragalna/Sicily/ Italy | The duration of the simulation is 4 h with 4 h supply duration |
| Campisano et al. (2019b) | Outlets, tanks, and outfalls | Hyperbolic law of the emitter | Tank | 1D | Consideration of household tanks in modelling of IWS with EPA-SWMM | Flow entering the tank is modelled with an equation from De Marchis et al. (2015) | – | – |
| Gullotta et al. (2021) | Outlets, tanks, and outfalls | Hyperbolic law of the emitter | Tank | 1D | Improvement of equity in IWS with users has household tanks | A novel methodology based on optimisation is developed to improve equitable supply | Northern Italy | 14 days simulation. The amount of water continuously released from the supply RSV was assumed equal to 70% of the total daily demand |

In particular, the challenges in the modelling are:

1. The complexity of the flow during the filling and emptying process.
2. Calibration and validation of the parameters required to model household tank inflow.
3. The detailed data required to build a hydraulic model.
4. The uncertainties with regard to consumers' behaviour (e.g., pipe size and connectivity at the consumer's end, the existence of household tank and/or private pump, demand pattern, etc.).
5. The practical application of macroscopic models without looking into the detail of the system behaviour.

Section 3 presented the studies addressing the challenges and the peculiarity of IWS which were mentioned in Section 3.1. It is clear from the above discussion that intermittency in the water supply affects the WDS's behaviour as well as consumers' behaviour. The following section (Section 4) discusses the issue of inequity in water distribution among consumers as a result of IWS, the current measures of equity, and the existing methodologies for equity improvement. It is important to note the critical role of IWS hydraulic models in the process of decision-making with regard to equity improvement interventions.

## 4. WATER DISTRIBUTION EQUITY

PDA is suitable for modelling the fully pressurised state of IWS as it provides a more realistic calculation of pressures and flows in IWS where pressures are usually insufficient to satisfy full demand. This way, DNs appear to have different levels of demand satisfaction and the inequity in the distribution of water is observed. The issue of inequity is evident in most of the demonstrated IWS hydraulic modelling case studies and it has been addressed in several other publications that do not use explicit hydraulic modelling. Current measures of supply equity as well as proposed interventions from the literature are discussed in the following subsections.

### 4.1. Measures of supply equity

Different measures of equity/inequity have been proposed. Two performance indices are used in Fontanazza *et al.* (2007) to evaluate the equity of water distribution: (i) the supply ratio (SR) between the water volume supplied to the consumers in a service cycle ($V_{sup}$) and the consumers' demand ($V_{dem}$): $V_{sup}/V_{dem}$ and (ii) the ratio between the water flow discharged to the user during a service day in intermittent ($Q_{int}$) and continuous distribution conditions ($Q_{cont}$): $Q_{int}/Q_{cont}$.

Ameyaw (2011) used a measure called deviation of equity and is defined as:

$$D_E = \text{Min} \sum_{i=1}^{n} |(\%Q_{av} - \%Q_s)| \tag{7}$$

where $N$ is the number of consumer nodes in the WDN. $Q_s$ (%) is the SR of each DN as a percentage (actual volume supplied to the amount required), and $Q_{av}$ (%) is the average ratio of water delivered to all DNs in the network as a percentage.

Chandapillai *et al.* (2012) measured how inequitable the distribution of an available quantity of water focusing on the minimum SR observed in the network is. The drivers of the optimisation procedure are the minimisation of cost and the increase of SR of the most disadvantaged/deficient DN and the respective index is shown in the equation subsequently:

$$\text{Inequity} = 1 - \text{Min} \left\{ \frac{V_{sup}}{V_{dem}} \right\}_j \tag{8}$$

where $V_{sup}$ is the volume of water supplied during the service cycle, $V_{dem}$ is the volume of demand during the service cycle, and $j$ is the DNs' number.

Authors note that optimisation of pipe diameter with these objectives could result in a network with very small diameter pipes which might not satisfy required flow rates. Hence, a minimum head requirement is also incorporated as a constraint: the minimum head surplus of the network DNs ($H_{avl} - H_{min}$) should be greater than zero.

Gottipati & Nanduri (2014) proposed a new equity assessment index, namely uniformity coefficient (UC), which considers the deviation of each node's SR from all nodes' average supply ratio (ASR). The deviation of the SR of the node from the ASR is computed at each node, and the mean of these deviations is defined as the average deviation of the sum of each DN supply ratio from the ASR (ADEV). UC is defined as:

$$UC = 1 - \left(\frac{ADEV}{ASR}\right) \tag{9}$$

Different scenarios were considered in the analysis of a real case study to assess the impact of the source elevation, the variation of demands at certain nodes, and changes in pipe diameters in the UC. Two sample networks are used to assess the impact of the location of the source RSV and the effect of the layout of the network in the UC.

In Vairavamoorthy (1994), the overall objective was to distribute the limited quantity of water of an IWS as fairly and equally as possible. Considering water-starved systems (Section 3.1) where pressure dictates the quantity of water collected by the consumers, the objective for the improvement of equity was set as minimisation of pressure diversity at the minimum cost. Ilaya-Ayza *et al.* (2017) followed a similar pressure-oriented approach for optimising operational scheduling, which aims to maintain pressures in the network close to some reference pressure.

Overall, either the volume SR or pressure uniformity is the core element of the current equity indices used in the hydraulic modelling of IWS during its fully pressurised state. The use of a pressure equity measure could be justified in cases where there is high uncertainty about demands or if it is known that consumers' behaviour resembles more closely flow through uncontrolled outlets. In this case, care should be taken that aggregated DNs do not have large elevation differences as this is going to affect the accuracy of results. For example, lower elevation aggregated nodes would, in reality, have higher available head pressures and thus larger supply ratios, and *vice versa*. A major shortcoming of the above hydraulic models that use the volume SR indices to measure equity is that they do not consider the filling process during which a large volume of water is delivered in an inequitable manner (e.g., nodes closer to the source start receiving water several minutes or hours before the furthest ones). New equity measures should be defined alongside improved models that address the above shortcomings. Again, the appropriate equity measure might differ according to each individual case study or hydraulic model used and such a choice should be well justified.

## 4.2. Equity improvement interventions

Various types of network interventions have been proposed in order to achieve a more equitable distribution of water. Vairavamoorthy (1994) and Batish (2003) considered the use of FCVs at specific pipes to constrain flow at sections that receive more water and increase flow at sections with low-flow rates. In Vairavamoorthy (1994), the problem was formulated with two aims/stages: first determine the optimum locations for the FCVs and secondly define the optimum valve settings for each. Gullotta *et al.* (2021) examined the installation of gate valves and valves of different levels of closure for improving equity. The pipes on which the valves are installed as well as the valves' level of closure were used as the decision variables in the optimisation problem. NSGA-II was used for the optimisation and the objective was the maximisation of UC (as defined in Gottipati & Nanduri (2014)) with the constraint of not reducing the SR at demand nodes that initially had an SR value of >0.70.

Chandapillai *et al.* (2012) considered the design phase of an IWS and the diameters of all pipes were used as the decision variables. Ameyaw (2011) and Gottipati & Nanduri (2014) considered the possibility of installing elevated storage tanks in the network for equity improvement. In Ameyaw (2011), a multi-objective optimisation algorithm is used to generate solutions with regard to the number and size of the elevated storage tanks. The objectives of the optimisation are maximisation of equity (deviation of equity, see the above subsection) of distribution among consumers and minimisation of cost.

In terms of operational scheduling, Ilaya-Ayza *et al.* (2017) considered different supply schedules for each subnetwork of a WDN in order to improve the conditions and quality of service for the consumers. The main goals of the multi-criteria optimisation process are to maintain pressures across the network close to a reference pressure, reduce inconvenience for consumers, modify the shorter supply schedules, and ease sector operation. Supply sessions are split into different supply blocks and are optimised for which subnetworks are supplied during each supply block.

Overall, the use of hydraulic elements of different attributes such as different types of valves with various settings, service tanks, and gate valves has been studied. All the above have proved to be efficient in improving water distribution equity across WDNs even if the applicability of such methodologies in large and complex networks is questionable. The vast search space for different solutions in terms of topology and setting (e.g., for valve closure) complicates the process of finding an optimal solution and increases the computational cost significantly. More research in the field of cost-efficient interventions and operational management of hydraulic elements is required for the improvement of water supply service, equity in distribution, water quality, and leakage reduction.

## 5. CONCLUSIONS AND RECOMMENDATIONS

Access to clean potable water should be ensured for all people even if, under current IWS practices, this necessity is not guaranteed in terms of quantity or quality. IWS is considered as a measure to cope with the inability of the system or lack of resources to provide CWS. In many cases, though, it has been established as the norm, and the water supply conditions seem to further deteriorate under the IWS regime. IWS should be avoided whenever possible, but in cases where it cannot, comprehensive design principles are needed. In cases where IWS is established, new management approaches are needed in order to improve the quality of service and push forward for converting the operation of the system to CWS. Interventions in the physical components or operational management of the system can provide significant improvement. The hydraulic modelling aspect is critical for working on the above. Understanding of conceptual behaviour (ranging from individual user behaviour to the entire system) of the system is at the heart of this issue.

In this study, first, fundamental WDN pressure-driven hydraulic analysis methods were covered to establish a foundation for the main subject of this paper. After that, a wide range of approaches to hydraulic analysis methods of IWS were covered. From the examined literature, it can be highlighted that the choice of hydraulic model is done for each case individually taking into account the specific case characteristics (i.e., availability of water, user behaviour, existence of household tanks, and pumps). It is established from observation in IWS that low pressures are dominant in the network. Hence, the pressure-dependent hydraulic model is required to be used for IWS. An important aspect that is usually neglected is the network filling process. EPANET is not able to simulate the filling process, but EPA-SWMM is capable of doing it. However, the software is for the urban drainage system, and tools need to be adjusted to represent a WDN. There are also other attempts for a more accurate simulation of the filling process, but they are in the early stage of commercial application.

With regard to the available simulation software reviewed above, the EPANET-based ones have the advantage of being able to implement network elements and features related to water supply systems, while EPA-SWMM lacks some of these features (e.g., different types of pumps and valves, or water quality modelling).

While all the above models are insightful for better understanding the behaviour of IWS systems, more robust and complete models are needed in order to support decision-making in system operation and intervention planning. The knowledge gap still does exist and there is a need for developing more appropriate tools and methodologies that holistically address the unique system characteristics of the water distribution systems in developing countries. A realistic model should be able to simulate partial flow during the filling and emptying processes or in the cases of air intrusion in the pipeline system due to low pressure.

To conclude, the hydraulic modelling of IWS systems is still an area under research. The current literature indicates that the models developed so far are not holistic in a way that they can be applied to real IWS. Further research should address the following key gaps in knowledge:

§ The behaviour of water users under pressure deficient conditions needs to be analysed and understood better with the aim of improving the characterisation and estimation of water consumption in IWS systems. This is vital for improving the existing IWS hydraulic models and tools as water consumption is the driving force of these systems.

§ Water losses, either from leaks or unauthorised connections, are a huge issue in IWS systems and are critical components of the modelling of IWS systems. It is well known that background losses increase after the establishment of IWS and these continue to increase since network components deteriorate rapidly. Macroscopic models like the one introduced by Taylor *et al.* (2019) can be beneficial for getting insight into the system's behaviour. Improved understanding and characterisation of water losses in IWS systems are critical for the hydraulic modelling of these systems.

§ The water filling process of an IWS system remains under research. Hydraulic models of IWS systems that consider the filling in and emptying out processes in addition to the pressurised state of the IWS need to be developed.

§ There is a need to establish procedures and systems for a more systematic collection of wide-ranging data in real IWS systems, so that hydraulic models of these systems could be more accurately built and calibrated. Information about IWS system topology and existing physical components is critical for its hydraulic modelling but often not available. Critical information is also often missing for the operation, maintenance, and performance of these systems. Collecting all this data will lead to better models with more accurate predictions and a clearer view of how equitable the water distribution is in reality.

Furthermore, once the above issues are addressed and improved hydraulic models of IWS systems are created, these could be used to improve the planning and operation of these systems by addressing a wide range of relevant issues such as improved equity of water supply via better scheduling and/or system configuration changes (including the creation of district-metered-areas), reduction of water losses, and improved water quality. Accordingly, a new hydraulic analysis method based on EPA-SWMM will be proposed in the subsequent study.

## ACKNOWLEDGEMENTS

The first author gratefully acknowledges the Republic of Turkey Ministry of National Education for funding the PhD scholarship. The second author acknowledges the scholarship and support received from the UKRI and EPSRC Centre for Doctoral Training in Water Informatics: Science and Engineering (WISE) EP/L016214/1 programme.

## DATA AVAILABILITY STATEMENT

All relevant data are included in the paper or its Supplementary Information.

## CONFLICT OF INTEREST

The authors declare there is no conflict.

## REFERENCES

Abu-Madi, M. & Trifunovic, N. 2013 Impacts of supply duration on the design and performance of intermittent water distribution systems in the West Bank. *Water International* **38** (3), 263–282. https://doi.org/10.1080/02508060.2013.794404.

Ameyaw, E. E. 2011 *Investigating Intermittent Water Supplies in Developing Countries Using EPANET2*. M.Sc. Thesis, University of Exeter, Exeter, UK.

Ameyaw, E. E., Memon, F. A. & Bicik, J. 2013 Improving equity in intermittent water supply systems. *Journal of Water Supply: Research and Technology – AQUA* **62** (8), 552–562. https://doi.org/10.2166/aqua.2013.065.

Arregui, F., Cabrera, E., Cobacho, R. & García-Serra, J. 2006 Reducing apparent losses caused by meters inaccuracies. *Water Practice and Technology* **1** (4), wpt2006093. https://doi.org/10.2166/wpt.2006.093.

Arregui, F. J., Soriano, J., García-Serra, J. & Cobacho, R. 2013 Proposal of a systematic methodology to estimate apparent losses due to water meter inaccuracies. *Water Supply* 1324–1330. https://doi.org/10.2166/ws.2013.138.

Batish, R. 2003 A new approach to the design of intermittent water supply networks. *World Water & Environmental Resources Congress* 1–11. https://doi.org/10.1532/HSF98.20081003.

Bhave, P. R. 1981 Node flow analysis distribution systems. *Transportation Engineering Journal of ASCE* **107** (4), 457–467.

Bhave, P. R. & Gupta, R. 2006. *Analysis of water distribution networks*. Alpha Science International, London, UK.

Butler, D. & Memon, F. 2006 *Water Demand Management*. IWA Publishing, London.

Cabrera-Bejar, J. A. & Tzatchkov, V. G. 2009 Inexpensive modeling of intermittent service water distribution networks. In *World Environmental and Water Resources Congress 2009*, pp. 1–10. https://doi.org/10.1061/41036(342)29.

Campisano, A., Gullotta, A. & Modica, C. 2019a Using EPA-SWMM to simulate intermittent water distribution systems. *Urban Water Journal* **15** (10), 925–933. https://doi.org/10.1080/1573062X.2019.1597379.

Campisano, A., Gullotta, A. & Modica, C. 2019b Modelling private tanks in intermittent water distribution systems by use of EPA- SWMM. In: *Proc., 17th Int. Computing & Control for the Water Industry Conf*, Exeter, UK. University of Exeter.

Chandapillai, J. 1991 Realistic simulation of water distribution system. *Journal of Transportation Engineering* **117** (2), 258–263.

Chandapillai, J., Sudheer, K. P. & Saseendran, S. 2012 Design of water distribution network for equitable supply. *Water Resources Management* **26** (2), 391–406. https://doi.org/10.1007/s11269-011-9923-x.

Charalambous, B. 2012 The effects of intermittent supply on water distribution networks. *Waterloss 2012 February 26–29* (June 2011), 1–8. https://www.leakssuitelibrary.com/wp-content/uploads/2019/08/Bambos-EU-Ref-Doc-Case-Study-update.pdf.

Charalambous, B. & Laspidou, C. 2017 *Dealing with the Complex Interrelation of Intermittent Supply and Water Losses*. IWA Publishing, London, UK.

Criminisi, A., Fontanazza, C. M., Freni, G. & La Loggia, G. 2009 Evaluation of the apparent losses caused by water meter under-registration in intermittent water supply. *Water Science and Technology* **60** (9), 2373–2382. https://doi.org/10.2166/wst.2009.423.

Danilenko, A., Van den Berg, C., Macheve, B. & Moffitt, L. J. 2014 *The IBNET Water Supply and Sanitation Blue Book 2014: The International Benchmarking Network for Water And Sanitation Utilities Databook*. World Bank Publications, Washington, DC.

De Marchis, M., Fontanazza, C. M., Freni, G., La Loggia, G., Napoli, E. & Notaro, V. 2010 A model of the filling process of an intermittent distribution network. *Urban Water Journal* **7** (6). https://doi.org/10.1080/1573062X.2010.519776.

De Marchis, M., Fontanazza, C. M., Freni, G., Loggia, G. L., Napoli, E. & Notaro, V. 2011 Analysis of the impact of intermittent distribution by modelling the network-filling process. *Journal of Hydroinformatics* **13** (3), 358–373. https://doi.org/10.2166/hydro.2010.026.

De Marchis, M., Fontanazza, C. M., Freni, G., La Loggia, G., Notaro, V. & Puleo, V. 2013 A mathematical model to evaluate apparent losses due to meter under-registration in intermittent water distribution networks. *Water Science and Technology: Water Supply* **13** (4), 914–923. https://doi.org/10.2166/ws.2013.076.

De Marchis, M., Milici, B. & Freni, G. 2015 Pressure-discharge law of local tanks connected to a water distribution network: experimental and mathematical results. *Water (Switzerland)* **7** (9), 4701–4723. https://doi.org/10.3390/w7094701.

Dighade, R. R., Kadu, M. S. & Pande, A. M. 2014 Challenges in water loss management of water distribution systems in developing countries. *International Journal of Innovative Research in Science, Engineering and Technology* **3** (6), 2319–8753.

Dubasik, F. B. 2017 *Planning for Intermittent Water Supply in Small Gravity-Fed Distribution Systems: Case Study in Rural Panama*. Michigan Technological University, Houghton, MI, USA.

Erickson, J. J. 2016 *The Effects of Intermittent Drinking Water Supply in Arraiján, Panama*. PhD Thesis, University of California, Berkeley.

Fontanazza, C. M., Freni, G. & La Loggia, G. 2007 Analysis of intermittent supply systems in water scarcity conditions and evaluation of the resource distribution equity indices. *WIT Transactions on Ecology and the Environment* **103**, 635–644. https://doi.org/10.2495/WRM070591.

Fontanazza Chiara, M., Notaro, V., Puleo, V. & Freni, G. 2015 The apparent losses due to metering errors: a proactive approach to predict losses and schedule maintenance. *Urban Water Journal* **12** (3), 229–239. https://doi.org/10.1080/1573062X.2014.882363.

Freni, G., Marchis, M. D. & Napoli, E. 2014 Implementation of pressure reduction valves in a dynamic water distribution numerical model to control the inequality in water supply. *Journal of Hydroinformatics* **16** (1), 207–217. https://doi.org/10.2166/hydro.2013.032.

Fujiwara, O. & Ganesharajah, T. 1993 Reliability assessment of water supply systems with storage and distribution networks. *Water Resources Research* **29** (8), 2917–2924. https://doi.org/10.1029/93WR00857.

Galaitsi, S. E., Russell, R., Bishara, A., Durant, J. L., Bogle, J. & Huber-Lee, A. 2016 Intermittent domestic water supply: a critical review and analysis of causal-consequential pathways. *Water (Switzerland)* **8** (7). https://doi.org/10.3390/w8070274.

Germanopoulos, G. 1985 A technical note on the inclusion of pressure dependent demand and leakage terms in water supply network models. *Civil Engineering Systems* **2** (3), 171–179. https://doi.org/10.1080/02630258508970401.

Gottipati, P. V. & Nanduri, U. V. 2014 Equity in water supply in intermittent water distribution networks. *Water and Environment Journal* **28** (4), 509–515.

Gullotta, A., Butler, D., Campisano, A., Creaco, E., Farmani, R. & Modica, C. 2021 Optimal location of valves to improve equity in intermittent water distribution systems. *Journal of Water Resources Planning and Management* **147** (5), 04021016. https://doi.org/10.1061/(asce)wr.1943-5452.0001370.

Gupta, R. & Bhave, P. R. 1996 Comparison of methods for predicting deficient-network performance. *Journal of Water Resources Planning and Management* **122** (3), 214–217.

Ilaya-Ayza, A. E., Benítez, J., Izquierdo, J. & Pérez-García, R. 2017 Multi-criteria optimization of supply schedules in intermittent water supply systems. *Journal of Computational and Applied Mathematics* **309**, 695–703. https://doi.org/10.1016/j.cam.2016.05.009.

Ingeduld, P., Pradhan, A., Svitak, Z. & Terrai, A. 2006 Modelling intermittent water supply systems with EPANET. *8th Annual Water Distribution Systems Analysis Symposium* (1), 1–8. https://doi.org/10.1061/40941(247)37.

Kabaasha, A. M. 2012 *Modelling Unsteady Flow Regimes Under Varying Operating Conditions in Water Distribution Networks*. M.Sc. Thesis, UNESCO-IHE Institute for Water Education, Delft, The Netherlands.

Kansal, M. L., Kumar, A. & Sharma, P. B. 1995 Reliability analysis of water distribution systems under uncertainty. *Reliability Engineering and System Safety* **50** (1), 51–59. https://doi.org/10.1016/0951-8320(95)00051-3.

Klingel, P. 2012 Technical causes and impacts of intermittent water distribution. *Water Science and Technology: Water Supply* **12** (4), 504–512. https://doi.org/10.2166/ws.2012.023.

Klingel, P. & Nestmann, F. 2013 From intermittent to continuous water distribution: a proposed conceptual approach and a case study of Béni Abbès (Algeria). *Urban Water Journal* **11** (3), 240–251. https://doi.org/10.1080/1573062X.2013.765493.

Kumpel, E. & Nelson, K. L. 2016 Intermittent water supply: prevalence, practice, and microbial water quality. *Environmental Science and Technology* **50** (2), 542–553. https://doi.org/10.1021/acs.est.5b03973.

Kumpel, E., Woelfle-Erskine, C., Ray, I. & Nelson, K. L. 2017 Measuring household consumption and waste in unmetered, intermittent piped water systems. *Water Resources Research* **53** (1), 302–315. https://doi.org/10.1002/2016WR019702.

Lieb, A. M. 2016 *Modeling and Optimization of Transients in Water Distribution Networks with Intermittent Supply*. University of California, Berkeley.

Liou, C. P. & Hunt, W. A. 1996 Filling of pipelines with undulating elevation profiles. *Journal of Hydraulic Engineering* **122** (10), 534–539.

Macke, S. & Batterman, A. 2001 *A Strategy to Reduce Technical Water Losses for Intermittent Water Supply Systems (Diplomarbeit)*, *Fachhochschule Nordostniedersachsen*. University of Applied Sciences, Suderburg, Germany. Available from: http://sdteffen.de/diplom/thesis.pdf.

Mohan, S. & Abhijith, G. R. 2020 Hydraulic analysis of intermittent water-distribution networks considering partial-flow regimes. *Journal of Water Resources Planning and Management* **146** (8), 1–15. https://doi.org/10.1061/(ASCE)WR.1943-5452.0001246.

Mohapatra, S., Sargaonkar, A. & Labhasetwar, P. K. 2014 Distribution network assessment using EPANET for intermittent and continuous water supply. *Water Resources Management* **28** (11), 3745–3759. https://doi.org/10.1007/s11269-014-0707-y.

Mutikanga, H. E. 2012 *Water Loss Management Tools and Methods for Developing Countries*. PhD thesis, Delft University of Technology, Netherlands.

Mutikanga, H. E., Sharma, S. K. & Vairavamoorthy, K. 2011 Assessment of apparent losses in urban water systems. *Water and Environment Journal* **25** (3), 327–335. https://doi.org/10.1111/j.1747-6593.2010.00225.x.

Nyende-Byakika, S., Ngirane-Katashaya, G. & Ndambuki, J. M. 2012 Modeling flow regime transition in intermittent water supply networks using the interface tracking method. *International Journal of Physical Sciences* **7** (2), 327–337. https://doi.org/10.5897/IJPS11.873.

Nyende-Byakika, S., Ndambuki, J. M. & Ngirane-Katashaya, G. 2013 Modelling of pressurised water supply networks that may exhibit transient low pressure – open channel flow conditions. *Water Practice and Technology* **8** (3–4), 503–514. https://doi.org/10.2166/wpt.2013.054.

Pathirana, A. 2010 EPANET2 desktop application for pressure driven demand modeling. In *Water Distribution Systems Analysis 2010 – Proceedings of the 12th International Conference, WDSA 2010*, pp. 65–74. https://doi.org/10.1061/41203(425)8.

Prescott, S. L. & Ulanicki, B. 2008 Improved control of pressure reducing valves in water distribution networks. *Journal of hydraulic engineering* **134** (1), 56–65.

Puleo, V. 2014 Real-time optimal control of water distribution systems: models and techniques, including intermittent supply conditions. https://doi.org/10.1016/j.proeng.2014.02.102.

Reddy, L. S. & Elango, K. 1989 Analysis of water distribution networks with head-dependent outlets. *Civil Engineering Systems* **6** (3), 102–110. https://doi.org/10.1080/02630258908970550.

Rossman, L. E. 2017 *Storm Water Management Model Reference Manual Volume II – Hydraulics*. Available from: https://www.epa.gov/water-research/storm-water-management- model-swmm.

Segura, J. L. A. 2006 *Use of Hydroinformatics Technologies for Real Time Water Quality Management and Operation of Distribution Networks*. Case Study of Villavicencio, Colombia.

Shirzad, A., Tabesh, M., Farmani, R. & Mohammadi, M. 2013 Pressure-discharge relations with application to head-driven simulation of water distribution networks. *Journal of Water Resources Planning and Management* **139** (6), 660–670. https://doi.org/10.1061/(asce)wr.1943-5452.0000305.

Shrestha, M. & Buchberger, S. G. 2012 Role of satellite water tanks in intermittent water supply system. In *World Environmental and Water Resources Congress 2012: Crossing Boundaries, Proceedings of the 2012 Congress (Yepes 2001)*, pp. 944–951. https://doi.org/10.1061/9780784412312.096.

Simukonda, K., Farmani, R. & Butler, D. 2018 Intermittent water supply systems: causal factors, problems and solution options. *Urban Water Journal* **15** (5), 488–500. https://doi.org/10.1080/1573062X.2018.1483522.

Tanyimboh Tiku, T., Tabesh, M. & Burrows, R. 2001 Appraisal of source head methods for calculating reliability of water distribution networks. *Journal of Water Resources Planning and Management* **127** (4), 206–213. https://doi.org/10.1061/(ASCE)0733-9496(2001)127:4(206).

Tanyimboh, T. T. & Templeman, A. B. 2010 Seamless pressure-deficient water distribution system model. *Proceedings of the Institution of Civil Engineers: Water Management* **163** (8), 389–396. https://doi.org/10.1680/wama.900013.

Taylor, D. D. J., Slocum, A. H. & Whittle, A. J. 2018 Analytical scaling relations to evaluate leakage and intrusion in intermittent water supply systems. *PLoS ONE* **13** (5), e0196887. https://doi. org/10.1371/journal.pone.0196887

Taylor, D. D., Slocum, A. H. & Whittle, A. J. 2019 Demand satisfaction as a framework for understanding intermittent water supply systems. *Water Resources Research* **55** (7), 5217–5237. https://doi.org/10.1029/2018WR024124.

Trifunović, N. & Abu-Madi, M. O. R. 1999 Demand modelling of networks with individual storage. In *WRPMD'99: Preparing for the 21st Century*, pp. 1–10.

Vairavamoorthy, K. 1994 *Water Distribution Networks: Design and Control*. PhD Thesis, Imperial College of Science, Technology and Medicine, London, UK.

Vairavamoorthy, K. & Elango, K. 2002 Guidelines for the design and control of intermittent water distribution systems. *Waterlines* **21** (1), 19–21. https://doi.org/10.3362/0262-8104.2002.041.

Vairavamoorthy, K., Gorantiwar, S. D. & Mohan, S. 2007 Intermittent water supply under water scarcity situations. *Water International* **32** (1), 121–132. https://doi.org/10.1080/02508060708691969.

Vasconcelos, J. & Wright, S. J. 2005 Applications and limitations of single-phase models to the description of the rapid filling pipe problem. *Journal of Water Management Modeling*. https://doi.org/10.14796/jwmm.r223-19.

Vermersch, M., Carteado, F., Rizzo, A., Johnson, E., Arregui, F. & Lambert, A. 2016 *Guidance Notes on Apparent Losses and Water Loss Reduction Planning*. LeaksSuit Library, UK.

Wagner, B. J. M., Shamir, U. & Marks, D. H. 1988 Water distribution reliability: simulation methods. *Journal of Water Resources Planning and Management* **114** (3), 276–294.

Walski, T. M., Chase, D. V., Savic, D. A., Grayman, W. M., Beckwith, S. & Koelle, E. 2003 *Advanced Water Distribution Modeling and Management*. Bentley Institute Press, Waterbury, CT.

Walski, T., Blakley, D., Evans, M. & Whitman, B. 2017 Verifying pressure dependent demand modeling. *Procedia Engineering* **186**, 364–371. https://doi.org/10.1016/j.proeng.2017.03.230.

Walski, T., Havard, M. & Yankelitis, B. 2018 Testing pressure dependent demand at low pressure. In *WDSA/CCWI Joint Conference Proceedings*, Vol. 1.

Walter, D., Mastaller, M. & Klingel, P. 2017 Accuracy of single-jet water meters during filling of the pipe network in intermittent water supply. *Urban Water Journal* **14** (10), 991–998. doi:10.1080/1573062X.2017.1301505.

Weston, S. L., Loubser, C., Jacobs, H. E. & Speight, V. 2022 Short-term impacts of the filling transition across elevations in intermittent water supply systems. *Urban Water Journal* 1-10. DOI: 10.1080/1573062X.2022.2075764.

Zhou, L., Liu, D. Y. & Ou, C. Q. 2011 Simulation of flow transients in a water filling pipe containing entrapped air pocket with VOF model. *Engineering Applications of Computational Fluid Mechanics* **5** (1), 127–140. https://doi.org/10.1080/19942060.2011.11015357.

First received 13 February 2022; accepted in revised form 24 October 2022. Available online 16 November 2022

doi: 10.2166/aqua.2022.149

# Intermittent water supply in Indian cities: considering the intermittency beyond demand and supply

Suchismita Satpathy and Rohit Jha *

Department of Humanities and Social Sciences, Birla Institute of Technology and Science Pilani, Hyderabad Campus, Jawahar Nagar, Kapra Mandal, Hyderabad 500 078, India
*Corresponding author. E-mail: p20170439@hyderabad.bits-pilani.ac.in

SS, 0000-0001-9553-7203; RJ, 0000-0002-6742-5906

## ABSTRACT

Intermittent water supply (IWS) is a typical characteristic of cities in developing countries like India. One of the factors responsible for IWS is unaccounted for water (UFW). Factors like increase in population, upward trends in water demand, water scarcity due to climate change, and asymmetric distribution of water resources are also equally important. However, social relations of water are poorly understood and camouflaged under technicalities associated with IWS. Thus, in this paper, we examine IWS in Indian mega cities and secondary cities with an ethno-economic framework by bringing the data together from various administrative sources like government agencies, allude to its parameters from logistical perspectives, e.g. distances, capacities, population strength, etc., and try to position the water issue with challenges associated with caste, class, gender, religion, region, and governance. The ethno-economic perspective is an attempt to not only complement but also supplement the scientific studies from other disciplines by understanding the real nature of demand and supply problems in urban water management. This paper demonstrates IWS as a multi-dimensional problem and stresses the human drivers of intermittency.

Key words: ethno-economic approach, equitable water distribution, Indian cities, intermittent water supply, unaccounted for water, water supply systems

## HIGHLIGHTS

- The paper assesses the intermittent water supply (IWS) situation in mega cities and secondary cities of India.
- It is considering the understanding of IWS beyond the supply and demand factors to the people who deal with the problem on a day-to-day basis.
- It uses a framework combining the ethnographic and economic aspects which are important for equitable urban water distribution in IWS situation.

GRAPHICAL ABSTRACT

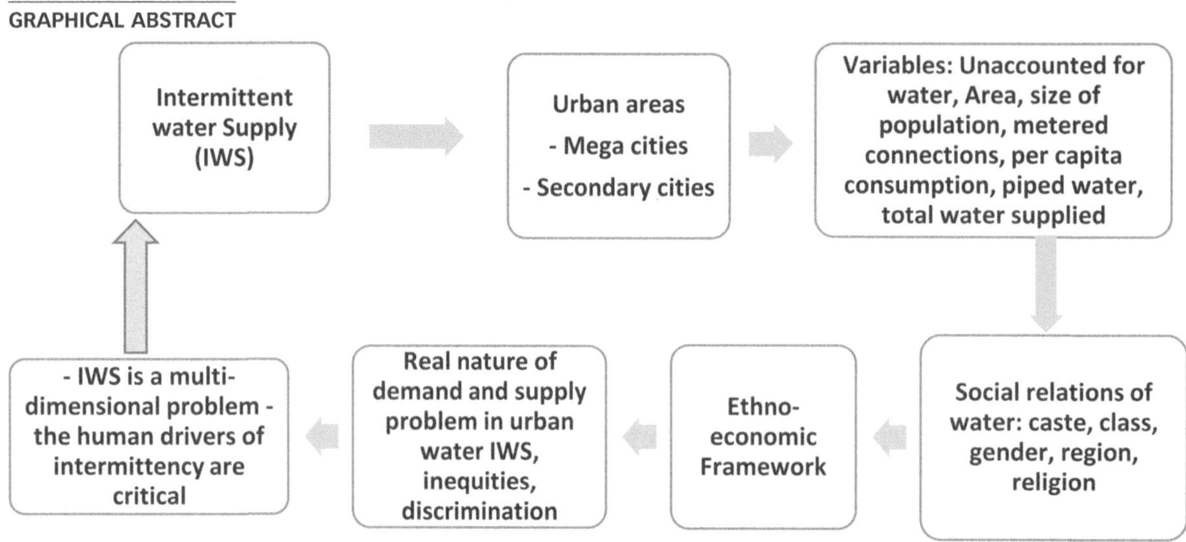


## INTRODUCTION

The sixth goal of the United Nations Sustainable Development Goals (SDGs) lays emphasis on providing clean drinking water and sanitation to all by 2030. However, the water sector is still struggling to improve water resource management and to increase the coverage and quality of water. The extent of coverage remains in question as millions continue to be out of formal networks of water supply in cities of Africa and Asia (Charalambous & Laspidou 2017). Along with coverage, intermittent water supply (IWS) is an additional challenge. The IWS system is defined as any piped system which provides water for less than 24 h in a day (Mokssit *et al.* 2018). Inadequate coverage, intermittent supplies, low pressure, water theft, poor infrastructure and poor quality are some of the most prominent features of water supply in Indian cities. The Indian urban water situation is such that 64% of the urban population is covered by individual connections and stand posts, duration of water supply ranges from 1 to 6 h, per capita supply of water in cities ranges from 37 Litres Per Capita Per Day (LPCD) to 298 LPCD for a limited duration, most cities do not have metering for residential water connections, 70% of water leakages are from pipes for consumer connection and due to malfunctioning of water meters, non-revenue water (NRW) or unac-counted for water (UFW) accounts for 50% of water production (Ahluwalia *et al.* 2011). With the rapid increase in urban population and continuing expansion of city limits, insufficient municipal budget, low cost recovery, etc. the challenge of deli-vering water in Indian cities is growing rapidly.

Though the commitment under SDGs 6 to 'leave no one behind' requires increased attention on disadvantaged groups and efforts to monitor elimination of inequalities in drinking water services, discriminations conspicuously exist. Categories of caste or class or community somehow determine where water does or does not go (Björkman 2015). The long-held belief that private participation can be a game changer by making public utilities more efficient and profitable has not yet materi-alized. On contrary, the situation in a few cities is such that the water mafia (Ranganathan 2014) and land mafia are collaborating as a nefarious nexus, thus adding new and unwelcome dimensions to the already existing water management problems. The inefficiency of the private sector is not only attributed to the failure of the age old water systems, but also to the intertwined discriminatory practices which show a wide range of intersectionality. For a private player, to navigate through these issues and ensure an equitable access to water seems impossible for now. Thus, apart from the supply-side economic issues, demand side population identities, awareness, and political power of the categories are also critical factors.

Various studies have identified the causes and the consequences of IWS worldwide (Klingel 2012; Ameyaw *et al.* 2013; Kumpel & Nelson 2014; Galaitsi *et al.* 2016; Mokssit *et al.* 2018; Simukonda *et al.* 2018; Ghorpade *et al.* 2021). The issue of equity emerges frequently in studies focusing on IWS (Galaitsi *et al.* 2016). The IWS results in inequitable supply, increased NRW, and deteriorated water quality (Ayoub & Malaeb 2006; Ghorpade *et al.* 2021). Insufficient metering, poor extent of coverage, leakages, water thefts, storage and distribution challenges, problems with access to water resources are responsible for IWS. UFW is one of the important causes and consequences of IWS. NRW or UFW is the water which does not make it from the source of the distribution network to the consumers due to leakages, wastage or theft. However, monitoring NRW and the differential impact of IWS on different sections of the population are problematic in an IWS regime (Al-Washali *et al.* 2019). Thus, understanding the problem of UFW is very much essential in the discourse of IWS. The present study aims to describe the existing intermittent urban water supply and distribution conditions of India's mega cities and secondary cities, linking them to the diverse impact this has on different sections of society. Using the ethno-economic framework, we attempt to further extend the understanding of IWS in social science.

Water Supply Systems (WSSs) are seen as a symbol of modernity and a way to have a better quality of life. However, in the context of UFW and IWS, modernity has failed and has resulted in commodification of water. The formal water networks not only challenged the notion of procuring water from streams, wells, rivers, and canals but also struggled to overcome the barriers of caste, class and religion to some extent. There are varied social actors and multiple power structures in the society that shape the nature of the social interactions in bureaucracy, community and among the people at large. Thus, the ethno-economic approach helps us interpret the situation of IWS better in the light of diverse social-economic categories.

In the following discussion, first we identify if there is any link between size of population, area of city and the existing intermittent water situation in the mega cities and secondary cities. The second part of the paper is the analysis of the social factors complicating the existing water situation in urban India and making it inequitable, inaccessible and scarce.

## INTERMITTENT URBAN WATER SUPPLY SITUATION AND UFW IN INDIAN MEGA CITIES

Water problems have already gripped the Indian mega cities (cities with population of 10 million or more) such as Mumbai, Delhi, Bengaluru, Kolkata, Hyderabad and Chennai. The secondary cities (population between 1–5 million) like Ahmedabad, Surat, Indore, Jaipur, Vadodara, Nagpur, Lucknow, Coimbatore, Nashik, Thiruvananthapuram, Varanasi and Visakhapatnam are not exempt from water issues. In India, where IWS is the norm, among the total connections available with all the water service providers only 43% of them are metered as of 2009 (Danilenko *et al.* 2014). The report by the Ministry of Urban Development indicates that in 2012, 13.3% of metered connections with 69.2 LPCD and 32.9% of NRW exist in the country against 100% metering, 132 LPCD and 20% NRW targets (Ghorpade *et al.* 2021). The average annual per capita water availability in the years 2001 and 2011 in India was assessed as 1,816 and 1,545 m$^3$, respectively, which may further reduce to 1,486 and 1,367 m$^3$ in the years 2021 and 2031, respectively (Ministry of Jal Shakti 2021). Only 47% of urban households have individual water connections in India, and about 40–50% of water is reportedly lost in the distribution system for various reasons. It is in this context that there is a need to examine the existing water systems and their social networks.

Table 1 contains the water profiling of the six mega cities. Over the decades, the population of Mumbai has grown enormously, giving rise to the sprawling slums within the city and suburbanization in the periphery, making it notorious for its congested living facilities. The situation for Delhi, and its periphery including Gurgaon and Faridabad which constitute the National Capital Territory, is complex due to the daily movement of a large number of people into and out from the city. This migratory population complicates the supply of all of the basic amenities, not just water. In part due to such complications, the current demand for water in the city is projected at 4,769 Million Litres per Day (MLD), whereas only 3,546 MLD is supplied. In Hyderabad, the HMWS&SB estimated the demand and supply of water at 2,312 and 1,343 MLD, respectively, for the year 2015. If the current trends of growth continue, then by 2021 the demand is projected to be at 3,172 MLD and supply at 2,214 MLD (Safe Water Network- Hyderabad 2016a). Thus, there is an obvious and large gap between the demand and the supply. This gap is closed by the means of exploiting groundwater. Groundwater in about 56% of the assessment units in Delhi was found to be overexploited. Its level is declining by 0.5–2 m annually in most parts of Delhi, and in Mumbai it is considered unfit for drinking. But the supply gap is met at household levels by accessing the groundwater (Safe Water Network -Delhi 2016b).

Kolkata, having a considerably high population, actually has the lowest quantity of water supplied out of all the mega cities with a considerable per capita consumption. The benchmark notified by the Ministry of Housing and Urban Affairs, Government of India is 135 LPCD for urban WSSs (2 March 2020). The per capita consumption for Kolkata is approximately 134 LPCD, which is close to the national average consumption and is further closer to the national goal of 150 LPCD for mega cities. However, it has been observed that the supply of water is erratic. In Chennai, per capita consumption of

**Table 1** | Water supply profile of the six mega cities

| City | Mumbai | Delhi | Kolkata | Chennai | Bengaluru | Hyderabad |
|---|---|---|---|---|---|---|
| Population | 1,84,14,288 | 1,67,87,941 | 1,41,12,536 | 86,96,010 | 84,99,399 | 77,49,334 |
| Area (km²) | 603.4 | 1,484 | 206.08 | 426 | 709 | 625 |
| Quantity of water supplied (MLD) | 3,350 | 3,546 | 599 | 830 | 970 | 1,287 |
| Ground water (%) | NA[a] | 9 | 11 | 9 | 40 | 30 |
| Piped water (%) | 76 | 81.3 | 92 | 98 | 93 | 70 |
| Per capita consumption (LPCD) | 135 | 274 | 134 | 90 | 100 | 162 |
| Percentage UFW | 20 | 52 | 35 | 40 | 40 | 40 |
| Metered connections (%) | 81 | 55 | 1 | NA[b] | 95.5 | 30 |
| Nodal Agency | The Hydraulic Engineer Department | Delhi Jal Board (DJB) | Kolkata Municipal Corporation (KMC) | Chennai Metropolitan water supply and sewerage Board | Bengaluru water supply and sewerage Board (BWSSB) | Hyderabad Metropolitan water supply and sewerage Board (HMWS&SB) |

*Source:* Census handbook 2011, websites of respective municipal authorities, CGWB report, TISS, Safe Water Network- Delhi Hyderabad 2016a.
[a]In Mumbai, though the percentage of ground water usage is not available, the GMMC reports 3,950 dug wells and 2,514 bore wells under operation for water supply purpose in the city.
[b]Metering water is not seen as commonly in Chennai as in Bengaluru, Mumbai or New Delhi.

water is 90 LPCD which is quite low. In November of 2019 Chennai Metro Water increased its water supply by about 100 MLD (Viswanath 2019). This benefit has not yet been received by the peripheral areas of Chennai which are heavily dependent on groundwater and informal sources of water like tankers.

Figure 1 reveals that Bangalore and Mumbai have done considerably well in metering of connections, thereby recovering the utility cost to a certain extent. In stark contrast, it appears that Delhi has one of the lowest percentages of metered

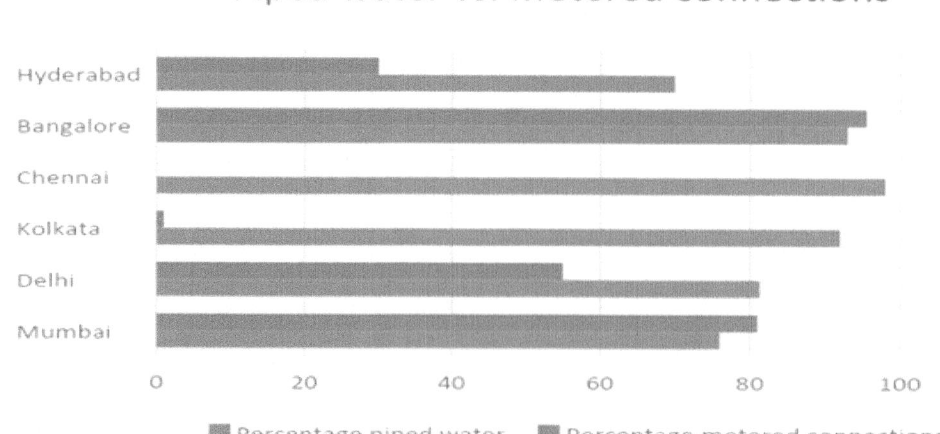


**Figure 1** | Piped water and metered connections.

connections among the mega cities. In 2011 the Delhi High Court exhorted in an order that there should be 100% metering in Delhi. Additionally, consumers should be asked to pay for the actual consumption. If citizens have a fundamental right to water, it is also important for them to pay for the water they use. But in reality, as much as 20% of the total 2 million connections in NCT of Delhi are unmetered. Most metered connections are also charged only a flat-rate tariff. Similarly, in Hyderabad, the piped water supply percentage stands at 70%, which is low compared to the other mega cities. There is a deficient piped supply of water to peripheral areas of the city, including the slums which mainly get their water supply through water tankers. Kolkata has an abysmally low number of metered connections, namely only 1% of households. Chennai recently started metering household and commercial water connections. This not only affects the utility's cost recovery but also gives scope for widespread wastage of water and UFW (Figure 2).

Total UFW is measured by the volume of the water lost (in litres) as a percentage or share of the total water supplied during the same time period. The tangible losses of existing intermittent supply in Indian cities are leaks and breaks in the pipes along with spill overs whereas the non-tangible losses are broken and tampered meters, inaccurate and improper readings, and outright water theft. The leaks and breaks which allow the water to seep through affect the quality of the water. UFW is the biggest problem in Delhi with almost 52% of water supplied being unaccounted for and this is not affordable where lesser quantities of water are supplied. Like most UFWs, the treated water supplied by the Delhi Jal Board (DJB) is lost during transmission and distribution due to pipeline leakage. The city also has a problem with illegal pilferage of water by tanker supply companies and illegal tapping of water. Piped water supply coverage in Delhi slums is 84.3% and of them, only 50.9% of the slum population have connections within their households, while the others share their source of drinking water.

In Mumbai, UFW accounts for 20%. It also boasts a higher percentage of metered connections at 81% and piped water supply coverage at 76%. The large slum-dwelling population causes a great amount of inequity in terms of distribution. Most slum residents have access to poor quality of water, inadequacies in the number of hours the water is supplied for, and so mostly resort to informal systems. The unaccounted-for water of about 20% comprises mainly pilferage and leakage from often 100-year-old pipes during transmission. The total water supplied is approximately 3,350 MLD and the unaccounted-for water is about 20% of this total, which is 670 MLD, a figure that can actually meet the demand and supply gap of water in the city to a large extent. Water supply through tankers is a vital support of water supply to the urban poor in Delhi. Over 800 tankers are owned and hired by DJB. But tanker water supply is quite costly for poor residents. There is always a risk of water getting contaminated, and it is difficult to purify it in order to make it drinkable. So, DJB in Delhi and VUDA in Visakhapatnam have started the treated-water kiosks throughout their cities to make drinking

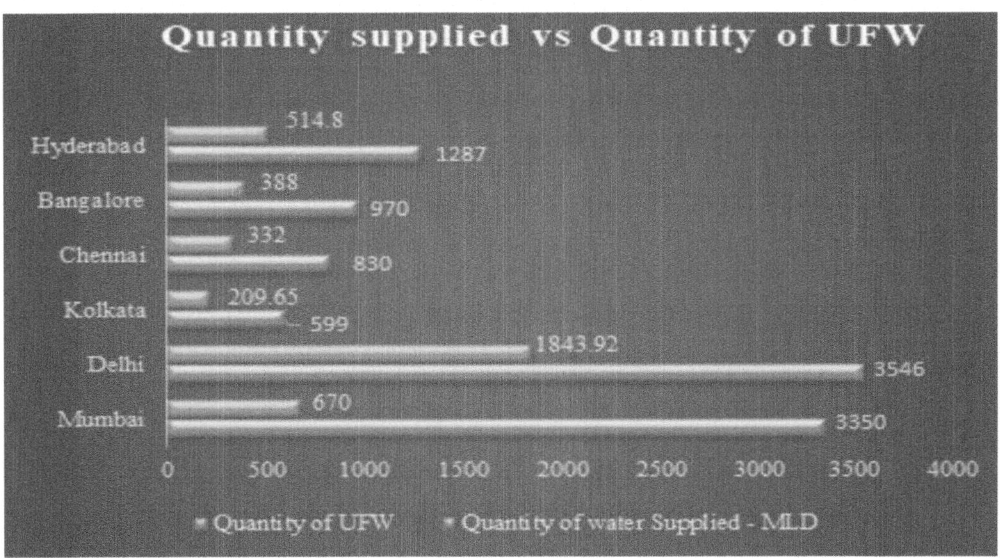


**Figure 2** | UFW vis-a-vis total water supplied.

water accessible for the urban poor. According to the BWSSB's Handbook of Statistics, in Bengaluru, 37–40% of the total water is lost in distribution itself, which may have improved due to the change in infrastructure, although the official latest numbers are not readily available. The high amount of unaccountable water has always plagued the BWSSB which incurs high losses (Mehta *et al.* 2013). Hyderabad faces about 20% of UFW issues mainly because of extensive and frequently occurring valve leakages, and under-maintained water supply infrastructure. Kolkata also has similar problems because of excessive distribution losses and poorly maintained pipelines.

Table 2 indicates the linear regression of the quantity of water supplied vis-à-vis variables like piped water, UFW and per capita consumption for mega cities. The relationship between the quantity of water supplied and piped water is strong with a correlation value of 0.994 and $R^2$ value of 0.988 and this indicates that an increase in the quantity of water supplied corresponds to an increase of piped water supply. The relationship between the quantity of water supplied and UFW is not so strong and this illustrates that an increase in the quantity of water supplied does not result in an increase in UFW. In this respect, the correlation is strong whereas the $R^2$ value of linear regression is not strong at 0.813 and 0.662. The per capita consumption of water and the quantity of water supplied exhibits a non-linear relationship with correlation value of 0.694 and $R^2$ value at 0.482, indicating that increased quantities of water need not lead to increase in per capita consumption. The second half of the table (Table 2) reveals the relationship between UFW and variables like piped water, area of the city, population and quantity of water supplied in six mega cities. On the basis of the $p$-values obtained for UFW vis-à-vis the piped water, water supplied and area, the relationship is statistically significant as the $p$-value is less than 0.05. An increase in values of piped water and quantity of water supplied does not necessarily result in the increase of UFW. However, in the context of area, any increase in the area shall cause an increase of UFW. The correlation determined between the population growth and the UFW has no inference as the $p$-value is higher than 0.05 and the correlation is statistically insignificant. This indicates that the growth in population does not necessarily contribute to an increase in the UFW. As per the dataset available, the effect of population growth on UFW is trivial.

## IWS SITUATION AND UFW IN INDIA'S SECONDARY CITIES

The secondary cities of today are the mega cities of tomorrow and the state of water infrastructure in these cities is a worry. An assessment of the water supply and distribution systems in selected secondary cities of India reveals that cities like Lucknow, Coimbatore, Visakhapatnam, Ahmedabad, Surat, Nagpur, etc. are also facing similar problems of urban water management, due to a rapid increase in population, resulting in the need for better urban planning, regulated growth, and decrease in environmental degradation (Ramachandra & Aithal 2013). Table 3 displays that Jaipur has almost 100% coverage in metering whereas Ahmedabad, Indore, and Lucknow have no metered connections. Metering of connections is a salient parameter to understand the patterns of usage of water, develop strategies to minimize the wastage of water, and assist in the financial recovery of the water utilities as consumers will be charged based on usage. It is to be noted that these cities have relatively higher percentages of piped water supply and their per capita consumption is more or less similar. Vishakhapatnam and Indore have lower piped water supply. Water tankers are used to provide water in Trivandrum's non-piped areas despite having 78% coverage of pipes and 100% metering. Nagpur supplies water to the urban residents at the cost of the irrigation potential as it is water-stressed and has significant metering.

Comparing the amount of water supplied to the city with the amount of water that is unaccounted for, we can see that the cities of Lucknow and Coimbatore have close to 55% water unaccounted for, which is extremely high. Most other cities have

**Table 2** | Correlation and linear regression of piped water, UFW; per capita consumption and area, population and UFW

| Mega cities (sample size: 6 cities): Independent variable – Quantity | | | Predicted variable: UFW | | | |
|---|---|---|---|---|---|---|
| Dependent variable | Correlation value | Linear regression: $R^2$ | Independent variable | Correlation value | $p$-Value | Linear regression $R^2$ |
| Piped Water | 0.994 | 0.988 | Piped water | 0.842 | 0.035 | 0.709 |
| UFW | 0.813 | 0.662 | Water supplied | 0.813 | 0.048 | 0.662 |
| Per capita | 0.694 | 0.482 | Area | 0.954 | 0.003 | 0.911 |
| | | | Population | 0.529 | 0.279 | 0.28 |

**Table 3** | Water situation in secondary cities

| City | Quantity of water Supplied – MLD | Percentage piped water | Per capita consumption – LPCD | Percentage UFW | Percentage metered connections |
|---|---|---|---|---|---|
| Ahmedabad | 1,210 | 90 | 140 | 23 | 0 |
| Surat | 980 | 95 | 140 | 15 | NA |
| Indore | 323 | 46 | 98 | 10 | 0 |
| Jaipur | 374 | 98.7 | 125 | 22 | 98 |
| Vadodara | 401.8 | 78 | 156 | 32 | 3 |
| Nagpur | 625 | 80 | 135 | 19 | 91.6 |
| Lucknow | 675 | 71 | 179 | 54.7 | 0 |
| Coimbatore | 137 | 88 | 125 | 56 | 30 |
| Nashik | 350 | 95 | 150 | 25 | NA |
| Thiruvananthapuram | 268 | 78 | 100 | 35 | 100 |
| Varanasi | 276 | 77 | 150 | 35.5 | NA |
| Vishakhapatnam | 291 | 54.9 | 113 | 30 | 2.16 |

*Source:* The data are collected from each city's municipal corporations.

close to the national average, i.e. 30%. High quantities of UFW are one of the biggest inefficiencies to plague the water utility system with dissatisfied consumers and poor infrastructure.

Table 4 reports the correlation and linear regression analysis of quantity supplied with the variables of piped water, UFW and per capita consumption of water in secondary cities. It highlights that the quantity of water indicates a strong relationship with piped water but a weak relationship with the UFW. The quantity of water increases with an increase in piped water and not necessarily with an increase in UFW. This observation in secondary cities is similar to the mega cities in India. Similarly, an increase in the quantity of water does not guarantee increased per capita consumption as the two variables have a poor linear relationship.

Similarly, the analysis of relationships between UFW and the other variables like piped water, area of the city and water supplied in secondary cities is predicted in Table 4. Observing the $R^2$ and correlation value of the three variables of piped water, water supplied and area, we can see that all three have a poor linear relationship with UFW and hence we cannot establish with certainty that a change in one of the variables can explain a corresponding change in UFW.

The examination of the intermittent water situation in urban India reveals that it is highly uneven and inequitable. Some cities like Meerut, Pune, Delhi, and Mumbai, which have the highest per capita water supply, appear to have access to a higher percentage of water than the percentage of the population that they account for (Malakar *et al.* 2018). But at the same time, these cities are among the most vulnerable to extreme water crises. Urban authorities often mention scarcity as the cause of IWS, but it is a multi-dimensional problem since intermittency is often human-driven. The urban water crisis is mediated by other aspects of the population in India like class, religion, region, gender, and caste. Thus, we link IWS with varied social categories.

**Table 4** | Correlation and linear regression of piped water, UFW and per capita consumption in secondary cities – UFW in secondary cities

| Secondary cities Independent variable: Quantity | | | Predicted variable: UFW | | UFW: Secondary cities (Sample size: 12 cities) | |
|---|---|---|---|---|---|---|
| Dependent variable | Correlation value | Linear regression $R^2$ | Independent variable | Correlation value | *p*-Value | Linear Regression $R^2$ value |
| Piped Water | 0.98 | 0.962 | Piped water | 0.602 | 0.038 | 0.362 |
| UFW | 0.674 | 0.455 | Water supplied | 0.674 | 0.016 | 0.455 |
| Per capita | 0.359 | 0.129 | Area | 0.355 | – | 0.126 |

## IWS AND THE SOCIAL CATEGORIES: HUMAN-DRIVEN INTERMITTENCY

Social equity in IWS is primarily about people, not water. Water may be allocated equitably, distributed equitably, and even accessed equitably, but if people are unable to derive benefits from it, the end result is not social equity. So, water equity should be judged by the final situation of people, and the distribution of the totality of benefits from water (Peña 2011). Irrespective of the region in an IWS situation, the consumers mostly secure their water supply through storing water during the time of supply. In such conditions, water quality is also affected as pipes are regularly drained and left without pressure between supply cycles, causing contamination. Nevertheless, intermittent supply and its effects on water quality can vary greatly between and within distribution networks as well. The problem of unequal distribution of water is more severe when there is IWS. So, the issue of equity in water supply is one of the major problems. However, equity in water supply is significantly affected by the location of the water tanks and the layout of the network (Gottipati & Nanduri 2014). Various empirical studies comment on how the access, quantity and quality of water vary when overlapped with caste, religion and income levels. The mitigation procedures undertaken by the households to overcome the scarcity and shortage of water are also linked to the socio-economic status. Though access to water is a basic human right and the key to achieve gender equality, and food security at the household level, water shortage continues to persist in the cities across the world. Water scarcity has affected the population of approximately 1.1 billion people. Intermittent hours of water supply force customers to rely on black markets (water mafia) or informal vendors, often serving higher-income citizens, thereby exacerbating inequalities among users and leading to commodification. The commodification of water affects the poor and the vulnerable groups in terms of housing, drinking water, sanitation and drainage facilities. IWS service costs more than regular service, and users bear the brunt of having to pay more to access water services via alternative routes (Charalambous & Laspidou 2017). It also weakens the social contract between governments and their communities when water utilities fail to deliver basic water services, perpetuating a downward spiral of water insecurity and fragility in many developing countries. We experienced this in South Africa in 2002. Riots broke out in Algeria in 2002, civil unrest erupted over use and allocation of water in China in 2000 and riots in Bolivia in 2000 were against privatization of the water system; conflicts over water shortages were reported in Chile, Lebanon, and Mexico in 2021.

### Class and IWS

The 69th round of the National Sample Survey (NSS) mentioned that in the year 2012, 94.1% of the households in the Indian slums had improved sources of drinking water. The report also highlights that the cost for procuring water is more for the poor and vulnerable groups as they are not connected to any grid systems or formal water supply networks. These sections of the population are dependent on private water suppliers and they end up paying more than required for the exact quantity of water they have access to. This is visible not only in Indian cities of Delhi but also in the UK where more than 3% of total expenditure to access water is termed as hardship. Empirical work on Asian and African cities illustrates that 11–20% of the total income is spent on access to water. The bargaining power and the political ideology also determine the ease of access to water. Thus, class as a social category does relate to the IWS.

Intermittency also roots a high risk of contamination leading to substantial health hazards. The consumers have to pay the costs, so called coping costs, for additional facilities, such as storage tanks, pumps, alternative water supplies, and household treatment facilities (Totsuka *et al.* 2004). The poor who cannot afford such facilities spend their time fetching water from public taps or vendors at comparatively high total costs.

In Mumbai, lower income households do not have adequate supply of water provided by the municipal corporation and have to resort to illegal measures, most commonly called water theft by the privileged sections of society (Graham *et al.* 2013). Similarly, in the city of Patna, many slum residents are still dependent upon informal sources of water like shallow wells to fulfill their demand for water (Sell 2013). However, the examples of Kerala and Rajasthan in India illustrate that informal settlements have better access to water. This intersectionality between water and class can be best seen in the urban and peri-urban areas of India (Das & Safini 2018) as exemplified by the inequality in access to water. In addition to poor accessibility, the low level of clean drinking water and a lack of information on healthy drinking water practices affect people's immunity and health. The upper- and middle-class community members get access to water by pressurizing and convincing the city officials to sanction water connections in the demanded areas. Migrants, on the other hand, were not a part of the original city plans, hence only receiving municipal services when political representatives perceive them as vote banks. Thus, access to water depends on the social relation to water that is based on the ability of residents to be recognized by city agencies through legitimate water services (Anand 2011, p.8).

IWS along with inequitable water distribution further worsens the status of the vulnerable class as the supply varies from 30 to 40 LPCD in low income neighbourhoods when compared to the 400 LPCD in high income neighbourhoods in Indian cities. On the one hand, there is implicit subsidization of tariffs for the rich through low water rates; on the other hand, there is inadequate coverage of water supply to the poor, hence forcing the urban poor to buy water at higher rates from private sources, turning the essential element for survival into an unaffordable luxury for them. The liberalized real estate market has further exacerbated the water demands. Thus, the main challenge according to Björkman 'is to make water flow to the unpredictable and constantly changing *location* of demand'. While policy perspectives tend to approach urban water questions with engineering and technological solutions, Björkman (2010) foregrounds the politics and power relations embedded in certain configurations of water access and flow across cities like Mumbai.

## Religion and IWS

Religious belief plays a significant role in shaping people's opinion about the environment (Jones 2014). Though water is considered precious by every religion, differential religious world views lead to over consumption, pollution, and scarcity. Furthermore, religion can also be a point of conflict in access to water among different religious communities where poor water access may be a deliberate state-led intent to discriminate against a particular religion (Mawani 2019). The 'Muslimness' of Mumbai slums complicates residents' access to water, a fairly commodified and politicized amenity; where water mafia plays an important role in making the state accessible, through the act of supervision and collaboration (Contractor 2012). Similarly, Muslim areas have poor access to municipal water in Ahmedabad. Water forms an essential part of most religious practices in almost all religions. Despite that, there is a paucity of literature on water and its relation to various religions. Today we have moved from sacred to secular water policy management. It has resulted in ignoring the intersectionality of religious sentiments with water (Priscoli 2012). Nevertheless, in India, Srilanka and South Africa, the intersectionality of religious philosophy with water management has produced some interesting case studies on awareness programmes about water conservation and management practices. But from the supply-side religious difference mediates decision-making and outcomes embodied by water distribution related technical plans. Thus, not only IWS but also religion and ethnicity influence the ability of communities to access water (Mawani 2019).

## Region and IWS

Intermittency generates inequitable water distribution due to pressure-dependent flow conditions, with obvious disadvantages for consumers located far away from the supply points or at higher altitudes in the area. The location and spatiality of the territories have influence on the appropriation, distribution and exploitation of the water resources. Nevertheless, the state is responsible for formal water supply and distribution networks to ensure everyone is provided with a fair quantity of water, which is safe and clean and in an equitable way. In India, despite focus on cities' water infrastructure, 24 × 7 continuous water supply in the cities is a distant dream and IWS further makes it inequitable given the uneven pressure conditions especially in the localities away from the reservoirs or the pumping stations. Those consumers farthest away from supply points will always collect less water than those nearer to the source. However, the physiographic constraints resulting in IWS are already discarded by technological developments. Additionally, regional or spatial unequal water access is believed to be human-driven (Anand 2011; Björkman 2015). The work of Anand (2011) highlights that provisioning of water relies on which sections of society demands, fights and receives the usage rights and not simply which altitude the consumer is in. This invokes the narratives of core-periphery, formality versus informality, and explains the 'fluid' nature of the territories and the access to natural resources associated with it.

## Gender and IWS

IWS affects women the most both in rural or urban areas as they are the primary users, providers and managers of water in households. Water is often associated with inherent femininity (Joshi 2011). So, men fetching water for the home is uncommon due to the traditional norms, stereotypes and fear of ridicule by other males (Crow & Sulthana 2002; Arouna & Dabbert 2010; Asaba *et al.* 2013).

Where intermittent access to clean water for drinking, hygiene and sanitation purposes exists the communities become more vulnerable to communicable diseases. Women's concerns in the water sector are articulated around their domestic roles and subsumed under notions of household and social equity. The larger questions of the water rights of women, both

**Table 5** | SC and ST population

Tap water from untreated source, all India data focussing SC and ST population

| S. no. | Locality | | Total in India | Total SC | Total ST | % SC from the total | % ST from the total | % SCST |
|---|---|---|---|---|---|---|---|---|
| 01. | Urban | Near the premises | 2,454,542 | 498,155 | 141,739 | 20.29 | 5.77 | 26.06 |
| | | Away from the premises | 638,625 | 137,990 | 49,558 | 21.60 | 7.76 | 29.36 |
| 02. | Rural | Near the premises | 10,566,863 | 2,365,760 | 1,118,870 | 22.38 | 10.58 | 32.97 |
| | | Away from the premises | 2,643,920 | 594,708 | 333,673 | 22.49 | 12.62 | 35.11 |

*Source:* Office of the Registrar General & Census Commissioner, 2011. https://censusindia.gov.in/nada/index.php/catalog/8480#metadata-description.

in terms of access and control over decision-making, remain unaddressed (Paul 2017). Women have to walk long distances in areas where water is not easily accessible and the fear of sexual assault, particularly at night, is a limitation to water access. So, the impact of differential supply and insecurity is more on women, especially from the poor and marginalized sections, illustrating that the women residents in slums and illegal settlements of Indian cities are affected most by the water problems (Das & Safini 2018, p. 191). Denial of water rights to women results in their deprivation of education, employment and social development. Much literature exists to back the gender-based inequalities in water management in rural areas of India, a similar if not more serious situation exists in the urban informal settlements. The sources of water in most informal Indian settlements like stand pipes, community hand pumps, wells and bore wells are inadequate, unreliable, and time consuming. Thus, the responsibility for the collection of water does in fact act like a roadblock to the wellbeing of women.

## Caste and IWS

Caste-based discrimination leading to human rights violations regarding drinking water makes it an important social issue. The hierarchical caste structure within the Indian society affects almost every aspect of life including the access to water. Given the intrinsic IWS system existing in India, caste-based water access complicates the water crisis further. The amount of time spent and the distance travelled for water collection has been consistently higher for the Scheduled Caste and Scheduled Tribe (henceforth SC & ST) communities when compared with other communities (Dutta *et al.* 2018). Access to water is a daily struggle for residents in rural and urban India due to the caste-based discrimination. Krishnaraj (2011) argues that, despite the policy initiatives and attendant programmes to expand access to water users, given our hierarchical society, the conversion of drinking water into a private good adversely affects women and the lower castes and classes.

Table 5 indicates the distance travelled by the residents of the SC and ST communities to fetch water from untreated sources which are either near or far from their homes. It clearly indicates that though the percentage of SC's and ST's accessing water from the untreated source away from premises in rural areas in India is higher than in urban areas, it is not significant when compared with the proportion of other castes in the rural area thereby busting the myth that discrimination with respect to distribution of water is only higher in rural areas.

## DISCUSSION

From the review of the IWS situation and social relations of water in India it is clear that the state is pivotal in the water space because it decides the mechanisms of appropriation, distribution and expropriation of water. The state provides access, provision, and distribution of drinking water, sanitation and drainage facilities. With liberalization and privatization, there is a shift in water governance. Though the state has a central role in the social relations of water, water governance has been problematized for over three decades now with changes in the role of the state in water provisioning post-liberalization.

The analysis of social relations and access to water in Indian cities have unpacked relationships that residents have with urban specialists. With water shortages increasingly cutting across class lines and legal statuses, access to municipal water supply in Indian cities is mediated by group identities like caste, gender, religion, space, etc. and actors such as state, political leadership and urban planners are playing very important roles. Thus, IWS is not simply an infrastructural problem but is partly human-driven where social factors are complicating the existing water situation.

Presenting an inquiry of this nature, the account of inequitable water access and severity of effects of IWS based on caste, class, gender and religion offered here is illustrative of how IWS seen as an infrastructural problem of scarcity and availability needs to be understood as a culmination of intense mediations between actors, identities, state and non-state actors with varying interests.

## LIMITATIONS

The study is mostly relying on the review of literature to substantiate the claim that social relations of water is crucial in IWS. Though it identifies the human drivers of intermittency, it is not providing any solution to the problem. An inquiry of this nature can only help understate the problem of IWS in a better way providing a fertile ground for future investigation on social drivers of IWS.

## CONCLUDING REMARKS

The question which still continues is if there exists a guarantee to provide safe water to the people in the developing countries by resolving the challenges of IWS systems or UFW. The focus of studies and the approaches of the state continue to be intensive towards creation of infrastructure in the form of the water supply networks which further perpetuate the challenges associated with inequality, power structures associated with caste, class and gender. This reinforces the central theme of Björkman and Nikhil Ananda's ethnographic works on the water situation in Mumbai which illustrate that access to water is beyond the economic dimension of water supply. Therefore, the need to merge the economic and ethnographic accounts is crucial to complement the technological interventions to improve the water accessibility and efficiency in the urban water distribution system. This ensures the fundamental right of access to water is upheld and the allocation of water resources and access to water facilities will be beneficial for all members of societies, regardless of their class, race, caste or sex.

## DATA AVAILABILITY STATEMENT

All relevant data are available from an online repository or repositories. Population finder | Government of India (censusindia.gov.in), Census tables | Government of India (censusindia.gov.in), India - HL-06 (SC): Households (excluding institutional households) from scheduled castes by main source of drinking water and location, India - 2011 (censusindia.gov.in), India - HL-06 (ST): Households (excluding institutional households) from scheduled tribes by main source of drinking water and location, India - 2011 (censusindia.gov.in), Greater Visakhapatnam Municipal Corporation (gvmc.gov.in), Home - Bangalore Water Supply and Sewerage Board (karnataka.gov.in), Home: Hyderabad Metropolitan Water Supply and Sewerage Board (hyderabadwater.gov.in), Official Website of Delhi Jal Board, Government of NCT of Delhi, India|, About Mumbai - MyBMC - Welcome to BMC's Website (mcgm.gov.in), Home page | Smartcities, Amdavad Municipal Corporation (ahmedabadcity.gov.in), Smart City Indore | Water Supply.

## CONFLICT OF INTEREST

The authors declare there is no conflict.

## REFERENCES

Ahluwalia, I. J., Munjee, N., Mor, N., Vijayanunni, M., Mankad, S. & Lall, R. 2011 *Report on Indian Urban Infrastructure and Services*. Ministry of Urban Development, Government of India, New Delhi.

Al-Washali, T., Sharma, S., Al-Nozaily, F., Haidera, M. & Kennedy, M. 2019 Monitoring nonrevenue water performance in intermittent supply. *Water* **11** (6), 1220.

Ameyaw, E. E., Memon, F. A. & Bicik, J. 2013 Improving equity in intermittent water supply systems. *Journal of Water Supply: Research and Technology – AQUA* **62** (8), 552–562.

Anand, N. 2011 Pressure: the polytechnics of water supply in Mumbai. *Cultural Anthropology* **26** (4), 542–564.

Arouna, A. & Dabbert, S. 2010 Determinants of domestic water use by rural households without access to private improved water sources in Benin: a seemingly unrelated Tobit approach. *Water Resources Management* **24** (7), 1381–1398. doi:10.1007/s11269-009-9504-4.

Asaba, R. B., Fagan, G. H., Kabonesa, C. & Mugumya, F. 2013 Beyond distance and time: gender and the burden of water collection in rural Uganda. *The Journal of Gender and Water* **2** (1), 31–38.

Ayoub, G. M. & Malaeb, L. 2006 Impact of intermittent water supply on water quality in Lebanon. *International Journal of Environment and Pollution* **26** (4), 379–397.

Björkman, L. 2010 Getting water in Mumbai: eight methods of access. In: *Western Political Science Association 2010 Annual Meeting Paper*.

Björkman, L. 2015 *Pipe Politics, Contested Waters: Embedded Infrastructures of Millennial Mumbai*. Duke University Press, Durham, NC.

Charalambous, B. & Laspidou, C. 2017 *Dealing with the Complex Interrelation of Intermittent Supply and Water Losses*. IWA Publishing, London, UK.

Contractor, Q. 2012 Quest for water. *Economic and Political Weekly [Preprint]*. Available from: https://www.epw.in/journal/2012/29/special-articles/quest-water.html

Crow, B. & Sulthana, F. 2002 Gender, class and access to water: three cases in a poor and crowded delta. *Society and Natural Resources* **15** (8), 709–724. doi: 10.1080/08941920290069308.

Danilenko, A., Van den Berg, C., Macheve, B. & Moffitt, L. J. 2014 *The IBNET Water Supply and Sanitation Blue Book 2014: The International Benchmarking Network for Water and Sanitation Utilities Databook*. World Bank Publications, Washington, DC.

Das, D. & Safini, H. 2018 Water insecurity in urban India: looking through a gendered lens on everyday urban living. *Environment and Urbanization Asia* **9**, 097542531878355. https://doi.org/10.1177/0975425318783550.

Dutta, S., Sinha, I. & Parashar, A. 2018 Dalit women and water: availability, access and discrimination in rural India. *Journal of Social Inclusion Studies* **4** (1), 62–79. https://doi.org/10.1177/2394481118774487.

Galaitsi, S. E., Russell, R., Bishara, A., Durant, J. L., Bogle, J. & Huber-Lee, A. 2016 Intermittent domestic water supply: a critical review and analysis of causal-consequential pathways. *Water* **8** (7), 274.

Ghorpade, A., Sinha, A. K. & Kalbar, P. P. 2021 Drivers for intermittent water supply in India: critical review and perspectives. *Frontiers in Water* **3**, 696630.

Gottipati, P. V. & Nanduri, U. V. 2014 Equity in water supply in intermittent water distribution networks. *Water and Environment Journal* **28** (4), 509–515.

Graham, S., Desai, R. & McFarlane, C. 2013 Water wars in Mumbai. *Public Culture* **25** (1), 115–141. https://doi.org/10.1215/08992363-1890486.

Jones, A. L. 2014 The impact of religious faith on attitudes to environmental issues and carbon capture and storage (CCS) technologies: a mixed methods study. *Technology in Society* **38**, 48–59.

Joshi, D. 2011 Caste, gender and the rhetoric of reform in India's drinking water sector. *Economic and Political Weekly* **46**, 56–63.

Klingel, P. 2012 Technical causes and impacts of intermittent water distribution. *Water Science & Technology Water Supply* **12**, 504–512. doi:10.2166/ws.2012.023.

Krishnaraj, M. 2011 Women and water: issues of gender, caste, class and institutions. *Economic and Political Weekly* **46** (18), 37–39.

Kumpel, E. & Nelson, K. L. 2014 Mechanisms affecting water quality in an intermittent piped water supply. *Environmental Science & Technology* **48**, 2766–2775. doi:10.1021/es405054u.

Malakar, K., Mishra, T. & Patwardhan, A. 2018 Inequality in water supply in India: an assessment using the Gini and Theil indices. *Environment Development and Sustainability* **20**. https://doi.org/10.1007/s10668-017-9913-0.

Mawani, V. 2019 Religiously unmapped water access: locating religious difference in access to municipal water supply. *Water* **11**, 1282.

Mehta, V. K., Goswami, R., Kemp-Benedict, E., Muddu, S. & Malghan, D. 2013 Social ecology of domestic water use in Bangalore. *Economic and Political Weekly* **48** (15), 40–50.

Ministry of Jal Shakti 2021 *Per Capita Availability of Water*. pib.gov.in. Available from: https://pib.gov.in/PressReleaseIframePage.aspx?PRID=1707522 (accessed 8 August 2022).

Mokssit, A., De Gouvello, B., Chazerain, A., Figuères, F. & Tassin, B. 2018 Building a methodology for assessing service quality under intermittent domestic water supply. *Water* **10** (9), 1164.

Paul, T. 2017 Viewing national water policies through a gendered lens. *Economic & Political Weekly* **52** (48), 77.

Peña, H. 2011 *Social Equity and Integrated Water Resources Management (No. 15)*. Global Water Partnership, Technical Committee (TEC), Stockholm.

Priscoli, J. 2012 Reflections on the nexus of politics, ethics, religion and contemporary water resources decisions. *Water Policy* **14**, 21. https://doi.org/10.2166/wp.2012.002.

Ramachandra, T. V. & Aithal, D. B. 2013 Urbanisation and Sprawl in the tier II city: metrics, dynamics and modelling using spatio-Temporal data. *International Journal of Remote Sensing* **3** (2), 66–75.

Ranganathan, M. 2014 'Mafias' in the waterscape: urban informality and everyday public authority in Bangalore. *Water Alternatives* **7** (1), 89–105.

Safe Water Network 2016a *Drinking Water Supply for Urban Poor: City of Hyderabad*. Available from: https://safewaternetwork.docsend.com/view/hjw5fhadmuzufin6

Safe Water Network 2016b *Drinking Water Supply for Urban Poor: City of Delhi*. Available from: https://safewaternetwork.docsend.com/view/jw8q24fh2byd37yr

Sell, S. 2013 Access to clean water: how Dalit communities in India are fighting for change. *The Guardian* 25 September. Available from: https://www.theguardian.com/global-development-professionals-network/2013/sep/25/indian-communities-fighting-for-change (accessed 3 August 2022).

Simukonda, K., Farmani, R. & Butler, D. 2018 Intermittent water supply systems: causal factors, problems and solution options. *Urban Water Journal* **15** (5), 488–500.

Totsuka, N., Trifunovic, N. & Vairavamoorthy, K. 2004 Intermittent urban water supply under water starving situations. In: *30th WEDC International Conference*, Vientiane, Lao PDR. Available from: https://core.ac.uk/download/pdf/288365164.pdf

Viswanath, M. 2019 Chennai's periphery yet to benefit from increased water supply. *The New Indian Express* 11 December. Available from: https://www.newindianexpress.com/cities/chennai/2019/nov/12/chennais-periphery-yet-to-benefit-from-increased-water-supply-2059958.html (accessed 3 August 2022).

First received 20 August 2022; accepted in revised form 16 November 2022. Available online 2 December 2022

doi: 10.2166/aqua.2022.024

# Hydraulic modeling approach for evaluating the performance of flow-starved water transmission networks

Abhishek Kumar Sinha **IWA** [a], Anujkumar Ghorpade **IWA** [a], Om Damani **IWA** [b] and Pradip P. Kalbar **IWA** [a],*

[a] Centre for Urban Science and Engineering, Indian Institute of Technology Bombay, Powai, Mumbai 400076, India
[b] Department of Computer Science and Engineering, Indian Institute of Technology Bombay, Powai, Mumbai 400076, India
*Corresponding author. E-mail: kalbar@iitb.ac.in

PPK, 0000-0002-1436-3609

## ABSTRACT

The planning and design of water networks in water supply systems are primarily based on demand-driven modeling. The prevailing design provisions, such as minimum diameter, lead to oversizing of the water network, affecting operation. In upstream balancing reservoirs, outflow due to the water transmission network's excessive withdrawal capacity surpasses the available inflow causing flow starvation under intermittent water supply. This flow starvation causes partial flow in the downstream vicinity and forms a standing water column in the balancing reservoir's immediate downstream pipe. Traditional modeling approaches cannot simulate the piped network performance under this phenomenon due to their inability to model partial flow. Hence, a novel modeling approach is developed using a tank with an irregular cross-section, which integrates the hydraulic performance of the tank and the downstream pipe. Additionally, a reservoir and control valve represent the water withdrawal mechanism at the downstream reservoir. The proposed modeling approach simulates performance of a flow-starved water transmission network. A case study based on a real network is used to illustrate the robustness of the proposed approach. The developed modeling approach can serve as a management tool to devise operation schedules, helping better manage the operations of the water networks.

Key words: flow starvation, hydraulic modeling, intermittent water supply, standing water column, water supply system

## HIGHLIGHTS

- The absence of effective demand-side management leads to flow starvation in water networks.
- Traditional models do not simulate flow starvation condition.
- A hydraulic modeling approach for a flow-starved water transmission network is developed.
- The study quantifies inequity in an intermittent water transmission network.

## GRAPHICAL ABSTRACT

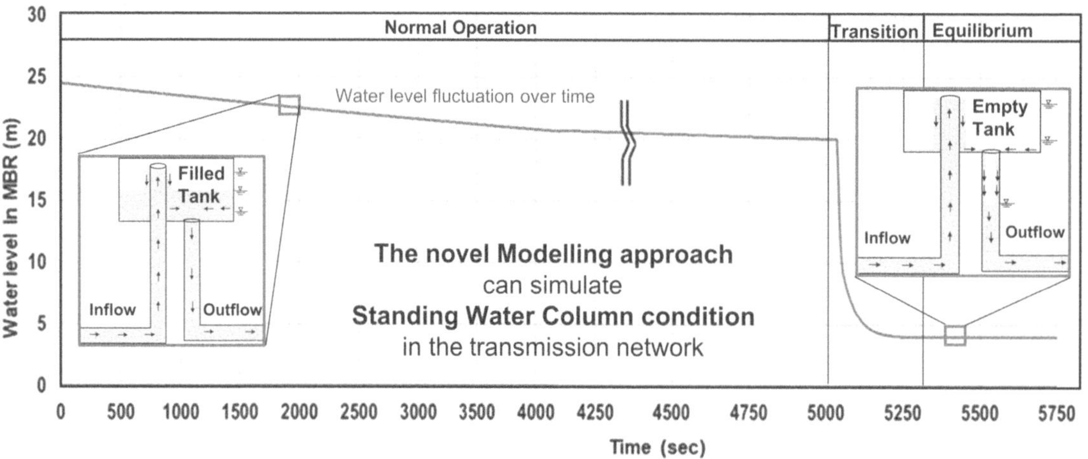

## ABBREVIATIONS

| | |
|---|---|
| CWS | continuous water supply |
| DDA | demand-driven analysis |
| EPS | extended period simulation |
| FDSR | flow-based demand satisfaction ratio |
| FSC | flow starvation condition |
| FSL | full supply level |
| GL | ground level |
| HGL | hydraulic grade line |
| IWS | intermittent water supply |
| LSL | low supply level |
| LPS | liter per second |
| MBR | master balancing reservoir |
| PDA | pressure-driven analysis |
| PDD | pressure-driven demand |
| PDO | pressure-driven outflow |
| PDA-FSC | PDA under FSC |
| SR | service reservoir |
| SWC | standing water column |
| VDSR | volume-based demand satisfaction ratio |
| WSS | water supply systems |
| WDN | water distribution network |
| WTN | water transmission network |

## NOTATIONS

| | |
|---|---|
| $H$ | height of FSL from the ground level of the MBR |
| $L$ | total length of the immediate downstream pipe |
| FSL | full-service level of the MBR |
| $L$ | total length of the immediate downstream pipe |
| LSL | low service level of the MBR |
| $W1$ | withdrawal capacity of the networks at source head equal to FSL of the MBR |
| $W2$ | withdrawal capacity of the networks at source head equal to LSL of the MBR |
| $W3$ | withdrawal capacity of the networks at source head equal to GL of the MBR |
| $h$ | height of LSL from ground level of MBR or length of the vertical portion of the immediate downstream pipe |
| $L\text{-}h$ | length of the horizontal section of the immediate downstream pipe |
| $x$ | distance of SWC from the LSL of the MBR at any arbitrary point in Stage 2 |

## INTRODUCTION

Despite intermittent water supply (IWS) not being ideal for supplying water to consumers, more than 1 billion people worldwide receive water intermittently (Charalambous & Laspidou 2017). Most of this population resides in developing countries (Charalambous & Laspidou 2017). Although IWS is discouraged due to contamination intrusion, consumer inconvenience, and the ill effect on infrastructure, its presence is a reality for a significant population (Charalambous & Laspidou 2017). Farmani *et al.* (2021) reaffirmed the presence of IWS in developing countries and assessed the resilience of IWS under COVID-19 conditions. Scarcity of water resources, unscientific network expansion, uncontrolled withdrawal, and high water loss are among the reasons for the prevalence of IWS (Simukonda *et al.* 2018; Kalbar & Gokhale 2019). Recently, Ghorpade *et al.* (2021) identified the drivers of the IWS as unsuitable design and analysis methods, lack of operation and maintenance, low consumer satisfaction, high non-revenue water, and weak institutional capacity.

Continuous water supply (CWS) provided to consumers with minimum threshold pressure, ensures equity as consumers are free to withdraw water as per demand (Randeniya *et al.* 2022). Achieving equity in water supply under IWS emerges as one of the challenges (Batish 2003; Randeniya *et al.* 2022). A precise operational control (pressure and demand) is needed to ensure that supply to consumers remains proportional to demand under IWS (Bhave & Gupta 2000; Batish 2003). Hence, a better prediction of flows in different parts of networks is vital for understanding the performance of IWS systems (Mohan & Abhijith 2020; Meyer & Ahadzadeh 2021). Recently, Sarisen *et al.* (2022) have examined and reviewed the challenges of modeling IWS systems and identified gaps in water demand modeling, leakage modeling, modeling of the water filling process, and the need for systematic data for real IWS systems.

## A brief review of the literature

The water network layout and its modeling are essential for understanding the performance of IWS systems. The following section discusses the typical layout of the IWS system adopted in India, followed by modeling studies. A water supply system (WSS) is a complex integrated system with multiple components, of which the water distribution network (WDN) is a dominant and complex component. A schematic diagram of a representative surface source-based WSS in India is presented in Figure 1. Water flows by gravity to a water treatment plant from an impounding surface water reservoir. After treatment, water is transported to the master balancing reservoir (MBR) through a pumping main. The MBR supplies bulk water through the water transmission network (WTN) to different service reservoirs (SRs), which deliver water through the WDN to consumers in different service areas (Ghorpade *et al.* 2021). The MBR and SRs are balancing reservoirs that differ in the pipe arrangement from their counterparts in the USA. In CWS systems, especially in developed countries, the floating tank is used, which has the advantages of better balancing behavior between inflow and outflow, water storage at nonpeak hours, and supporting pumps at peak hours (Walski *et al.* 2003; AWWA 2008). However, in India, inlet and outlet pipes are connected at full supply level (FSL) and low supply level (LSL), respectively, as shown in Figure 1 (Ghorpade *et al.* 2021). Adopting a floating tank in India could lead to colossal leakage and emptying of the tank, through the inlet (connection at LSL) in case of bursts and a leaky upstream network. Hence, despite the advantages, the floating tank is not feasible in India.

The WDN is used to supply water to consumers within service areas. In contrast, the WTN transports bulk water from the water source to SRs (Figure 1). Hydraulic modeling of WDN and WTN is used to understand the performance of WSS. Despite the inherent difference in purpose and layout, the same modeling approach is adopted for both network types. Various modeling and simulation software packages have been developed to predict the performance of water networks (Cembrowicz & Ates, 1998; Rossman 2000; Conety Ravi *et al.* 2019; Bentley 2022). There are two prominent analysis approaches; demand-driven analysis (DDA) and pressure-driven analysis (PDA). In DDA, nodal demands are presumed to

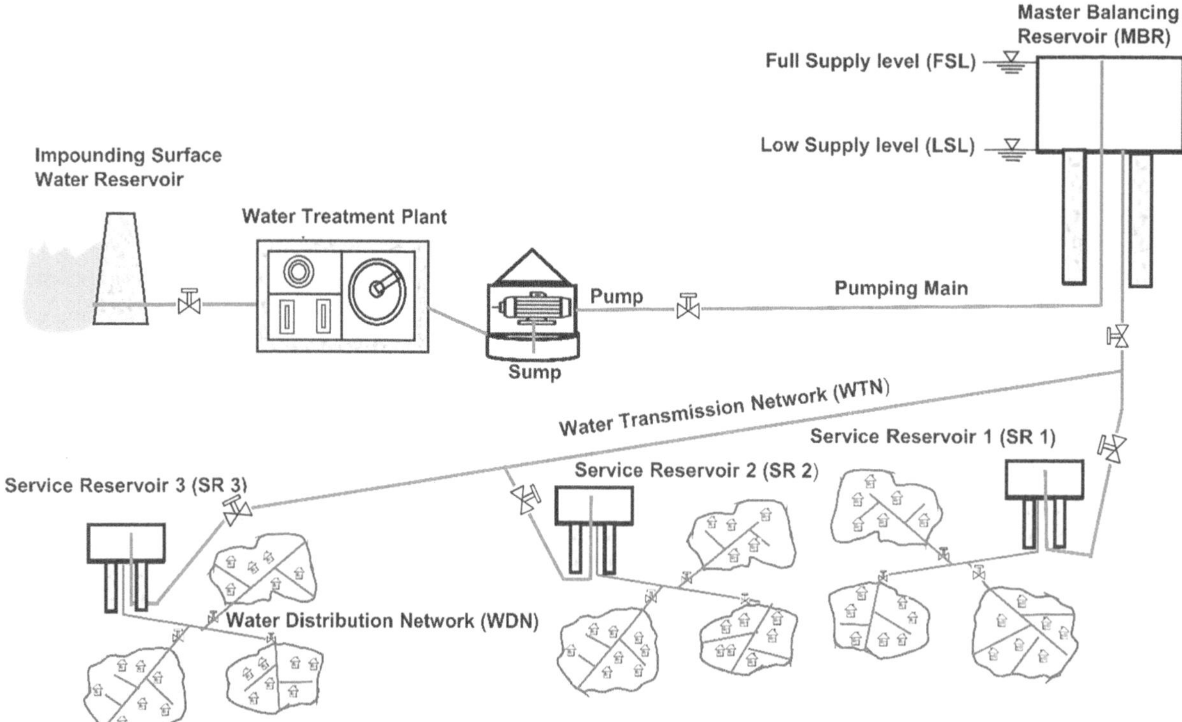


**Figure 1** | Typical layout of the WSS in India.

be completely satisfied irrespective of residual pressure available at these nodes (Gupta & Bhave 1996; Conety Ravi *et al.* 2019). In a situation of pressure deficiency, demand satisfaction at these nodes is impossible. Hence the DDA fails to predict the actual performance of WDN as nodal outflows are non-realistic in such pressure-deficient conditions. Bhave (1981) recognized this limitation and introduced the methodology of node-flow analysis, which later became the basis of PDA. Contrary to the DDA approach, the PDA approach is more realistic and uses the nodal head-discharge relationship for each node while analyzing the pressure-deficient WDNs.

Most PDA approaches implement demand modification based on outlet boundary condition, i.e. relating outlet pressure to flow (Bhave 1981; Wagner *et al.* 1988; Fujiwara & Li 1998), which is termed as PDA based on pressure-dependent demand (PDD). A few studies directly used outlet properties to determine the flow/outflow, termed as PDA based on pressure-dependent outflow (PDO) (Reddy & Elango 1989; Chandapillai 1991). Recently, Conety Ravi *et al.* (2019) reviewed and summarized developments in PDA studies. Other approaches, such as volume-driven analysis (Taylor *et al.* 2019; Sivakumar *et al.* 2020) are also based on these PDA modeling approaches and the similar water withdrawal mechanism.

Although PDAs have more realistic demand modeling of outlet boundary than DDA in pressure-deficient conditions (Gupta & Bhave 1996), most PDA studies model tanks as reservoirs, assuming infinite water sources, ignoring the inlet boundary condition. Whenever the upstream network fails to provide sufficient flow to match the flow prevalent in the downstream network, the downstream network operates at a lower flow; this phenomenon is termed herein as flow starvation. PDA studies predict a safer picture than the actual condition in such cases. Traditionally PDA models have focused solely on outlet boundary conditions, flow starvation requires more explicit modeling of inlet boundary condition. This limitation also arises due to modeling a single component (WDN or WTN without an upstream network), usually adopted in PDA, which conceals the effect of one component from the other. As a WSS is a highly integrated system, combined modeling of different components is necessary. A modified PDA approach is needed to account for flow starvation (a common phenomenon in IWS) to estimate flow and pressure accurately through the water network (Kalbar *et al.* 2014).

The modeling studies on IWS illustrated increased total outflow from source reservoirs and significant discharge variation at individual nodes (Tabesh *et al.* 2001). Reddy & Elango (1991) analyzed WDN performance under DDA and PDA in the IWS context and highlighted the difference in the flows. Specifically, Bhave & Gupta (2000) evaluated the performance of the WTN under PDA and show that the individual flow toward SRs significantly differ from DDA and results in increased total outflow from MBR. Furthermore, if the MBR fails to match this increased total outflow, the performance of the WTN is affected and the resulting flows deviate from PDA. Since the upstream network does not support the increased total outflow due to reasons such as limited pumping capacity, the total outflow from MBR through WTN will reduce, and individual flow toward the SRs will be different from those predicted by current PDA approaches. This change in performance is precisely what happens in the IWS amidst constant inflow into the MBR, as the inlet pipe discharges water into the atmosphere at the FSL, as shown in Figure 2(a). Here, the upstream inflow network fails to match increasing outflow; the total outflow decreases by reducing the available head, as MBR provides limited balancing storage. The total piezometric head exerted

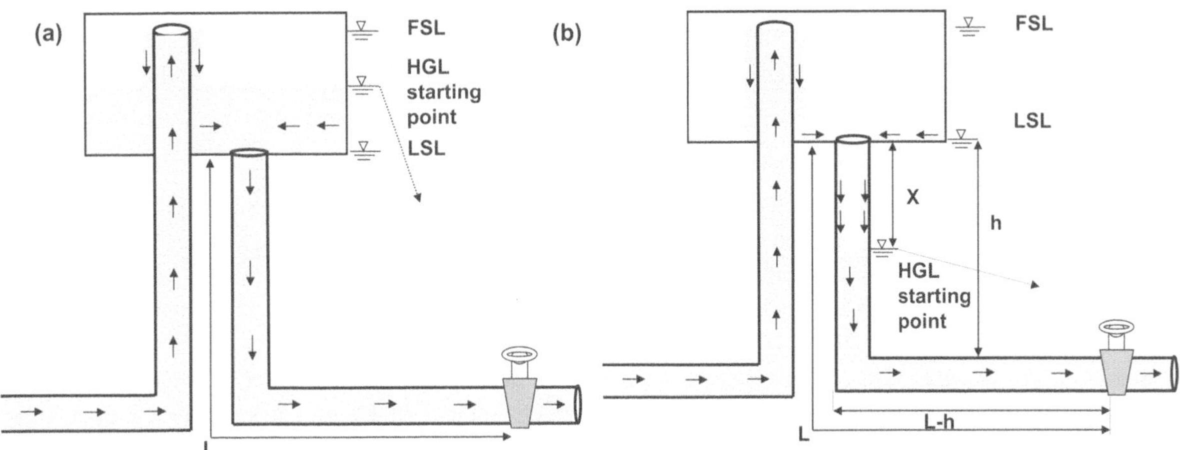


**Figure 2** | The water levels in MBR operation in (a) Stage 1 and (b) Stage 2 or 3.

on the WTN, represented by hydraulic grade line (HGL), depends on inflow. Hence in case of insufficient inflow, the piezometric head may go lower than the LSL of the MBR, as shown in Figure 2(b). This phenomenon is termed herein flow-starved condition (FSC). Kalbar & Gokhale (2019) have termed this condition as standing water column (SWC) and indicated that the network's excessive withdrawal capacity is the primary reason for SWC (Kalbar & Gokhale 2019).

### Need for the modified PDA approach

Various researchers have developed PDA models to simulate the IWS regime (Tabesh *et al.* 2001; Batish 2003; Ingeduld *et al.* 2007). Since conventional modeling approaches do not simulate partial empty pipe networks caused by pressure deficiency under flow starvation (Mohan & Abhijith 2020), the network's performance under FSC cannot be simulated. PDA models found in the literature do not consider the FSC prevalent in IWS (Mahmoud *et al.* 2017; Shirzad 2020). Modeling studies focused on IWS also fails to capture FSC (Mohapatra *et al.* 2014; Taylor *et al.* 2019; Randeniya *et al.* 2022). Mohan & Abhijith (2020) presented a novel modeling approach for partial flows under FSC; however, it is reported that mass-conservation is not maintained in their method (Meyer & Ahadzadeh 2021).

Additionally, the method's applicability is limited by two assumptions. Firstly, the model assumes fill-draw operation (drawing of water occurs subsequently after filling the tank). However, WDN with limited operational hours (2–3 h) can satisfy the fill-draw operation condition where the filling of the tank and supply from the tank does not overlap, but the same cannot be replicated in the WTN due to longer operational hours (16–24 h). Secondly, the model assumes the possibility of partial flow in any pipe within a WDN. This assumption is not valid in the WTN, where water is drawn from the MBR to the elevated SRs through vertically erected inlets, and hence the partial flow is not physically possible. Such an inlet can only carry flow when fully pressurized; therefore, it is not realistic to think of partial flow into the elevated SR. From this understanding, it is physically possible for water from the MBR, which is at a higher elevation, to fill a SR at a lower elevation as long as the water column has a high enough head; otherwise, no flow occurs into the SR.

Current modeling approaches do not account for the inlet boundary condition of WTN during flow starvation. Hence, the present study is focused on developing a hydraulic modeling approach for flow-starved WTN. The proposed approach is based on modeling WTN and presents a more realistic representation of the network and deployed control mechanisms. The present study associates partial flow in non-pressurized pipe in the downstream vicinity of balancing reservoirs with the inability of the upstream network to provide sufficient inflow and modify inlet boundary conditions accordingly.

## ORIGIN AND DRIVERS OF FLOW-STARVED CONDITIONS

### Typical WSS operation in India

WSSs in India are designed to operate as CWS systems as mandated by CPHEEO (1999); however, most systems operate as IWS. In IWS, most consumers do not receive the entire demand, and supply is uncertain (Farmani *et al.* 2021). Typically consumers store water in storage tanks to cope with intermittency (Agrawal *et al.* 2007) and start to draw water as soon as the supply is resumed (Ghorpade *et al.* 2021). The poor institutional capacity of water utilities cannot enforce demand-side management such as effective metering, appropriate tariff structure, action on non-payment (Burt & Ray 2014). Additionally, there are no proper valve controls on the water network to limit withdrawal except for complete shutoff, and zoning is done to limit withdrawal (Ghorpade *et al.* 2021). These operation practices result in the emptying of SRs due to over withdrawal. The empty SRs lead to unequal water distribution in downstream service areas and more withdrawal on upstream (Sashikumar *et al.* 2003; Kalbar *et al.* 2014).

### Drivers for FSC

The gap between the design and operation of WSSs in India is responsible for forming SWC, which is primarily caused by the inability to provide sufficient inflow, control withdrawal, or provide adequate balancing storage at any water tank. WSSs in India are designed with a design life of 30 years. The population is forecasted for the following three phases: (a) present phase: planning period; (b) intermediate phase: the period after 15 years from commissioning; and (c) ultimate phase: The period after 30 years from commissioning (CPHEEO 1999). Most IWS systems have the low capacity in terms of providing pressure as they fail to supply water with sufficient pressure due to simultaneous demand (Sashikumar *et al.* 2003; Kalbar *et al.* 2014). However, studies (Bhave & Gupta 2000; Tabesh *et al.* 2001) have shown that IWS leads to more water withdrawal than estimated demand. Field studies on IWS have reported an increase in peak factor indicating enhanced withdrawal (Sashikumar *et al.* 2003; Kalbar *et al.* 2014). This excess water withdrawal capacity is problematic. The WTN is designed for the ultimate phase, and the pump station is usually designed for the intermediate phase, which leads to higher withdrawal capacity of the WTN with respect to the upstream network (CPHEEO 1999).

Additionally, the provision of design norms such as the minimum diameter of the pipeline leads to higher design diameters than the actual requirement. This also increases the withdrawal capacity of WTNs significantly (Ghorpade *et al.* 2021). The WTNs have excess capacity for initial phase of design life (CPHEEO 1999). Hence, design reports recommend that the design flow be realized in the field with the help of valve throttling, head-dissipating devices (Bhave & Gupta 2000), and hydraulic isolating structures (Kalbar & Gokhale 2019). However, in practice, flow control is attempted with only sluice valves, resulting in poor performance (Ghorpade *et al.* 2021). Eventually, there is no valve operation over the period, and the WTNs operate with limited control in the poorly managed system. The tank configuration of balancing reservoirs, uncontrolled operations, and significant physical losses in the systems contribute to the SWC formation.

## The performance of a flow-starved system

The performance of a WSS under FSC is very different from that in CWS. Considering the tank configuration used in India, starting with an initially full tank, three different operation stages occur in the IWS system. These supply stages are described in the following section in the sequence of their occurrence (Figure 2).

### STAGE 1 (Normal): The water level is between FSL and LSL

In intermittent operation in India, inflow to the MBR is lower than outflow due to the WTN's higher withdrawal capacity. The volume of water in the MBR depletes at the rate of the difference between outflow and inflow. In Stage 1, inflow remains constant, whereas outflow decreases with a decrease in the water level. The system's performance in this stage is illustrated in Figure 2(a). The performance at this stage can be simulated in common software that allows PDA modeling using a tank element as a frictional resistance-free hydraulic element (Rossman 2000; Bentley 2022).

### STAGE 2 (Transition): The water level is below LSL

After Stage 1, transition starts, where vertical partial flow exists in the upper portion of the vertical pipe and the lower portion remains pressurized. During this transition, the water level in the MBR is below LSL (suppose at '*x*' distance from the LSL of the MBR); hence the pipe section immediately downstream to the MBR does not get fully pressurized (Figure 2(b)). During Stage 2, the WTN's withdrawal capacity changes with the pressurized part of the WTN. The decrease in water level leads to a reduction in the head (increase in *x*) and a reduction in the frictional resistance (reduction in the portion of pressurized pipe network). The immediate downstream pipe is vertical in most cases. Since the vertical pipeline is minimal in length with the least frictional resistant part of the WTN due to its higher diameter, the effect of reduced head in determining flow is more drastic than the reduced frictional resistance. As a result, the piped WTN's withdrawal capacity decreases, lowering the outflow.

### STAGE 3 (Equilibrium): Water level is below LSL and at equilibrium

After Stage 2, outflow decreases to match inflow, and equilibrium of water level is achieved in the pipe immediately downstream to the MBR. The equilibrium water level marks the beginning of Stage 3 and the end of Stage 2. At this stage, a unique combination of the head available to the WTN and its resistance is attained so that outflow of MBR equals inflow. The water level at this stage is depicted in Figure 2(b) when the operating water column becomes stable. This equilibrium stage is the SWC condition (Kalbar & Gokhale 2019).

After supply duration is over and the WTN is closed, the MBR is refilled in the remaining duration of the day so that the next day's supply can resume from Stage 1.

## METHODOLOGY

### Conceptual framework of the proposed modeling approach

This study develops a novel modeling approach predicting flows and pressure during FSC. This study proposes withdrawal capacity as an indicator of network capacity. The withdrawal capacity of the pipe system (pipe and control equipment) is defined with reference to the source water level. It is assumed that withdrawal capacity equals total outflow from an infinite volume water source at the constant head to a piped system under pressure-driven flow. In simpler terms, it is the total outflow expected through a piped system if a reservoir with infinite volume is connected to it under pressure-driven flow. Flow starvation occurs whenever inflow at any reservoir (with limited storage capacity) is lower than the piped network's withdrawal capacity and flow mismatch leads to a water level below the LSL of the reservoir. Hence, we propose a novel PDA under FSC (PDA-FSC) approach to model this condition.

In a flow-starved network, the pipe downstream to the storage reservoir behaves like storage reservoir and pipe. Although these storage reservoirs, i.e. MBR/SR are made of real/physical material having some hydraulic resistance, the reservoir has a larger diameter than the pipe network and negligible hydraulic resistance. Hence, the MBR/SR is modeled using a tank element with zero hydraulic resistance. The same approach can be extended to the vertical section of the immediate downstream pipe to the reservoir by considering it as part of the reservoir. Its resistance can also be neglected compared to the rest of the water network. For a more realistic representation of reservoirs under FSC, an irregular-shaped water tank is used instead of the conventional circular tank. The following assumptions are made and they provide the basis for adopting the proposed modeling approaches for the WTN under FSC.

*Assumption I*: MBR, along with the vertical section of the immediate downstream pipe modeled as a single tank element, can exhibit the same performance as physical MBR under FSC.

The vertical portion of the immediate downstream pipe's hydraulic resistance is assumed insignificant compared to the whole WTN. This allows us to model the MBR and the vertical outlet pipe as one hydraulic element. It is represented (Figure 3) as an irregular shape tank using tank element with LSL at Ground Level (GL) of the MBR without modifying the FSL. Volume and depth are used to obtain the cross-section curve of the tank element required to model irregular-shaped tanks in WaterGEMS Version 10 (Connect edition).

*Assumption II*: Physical WTN can be effectively represented in the hydraulic model WTN by excluding the vertical portion of the immediate downstream pipe to the MBR.

Assuming that the vertical portion of the immediate downstream pipe's hydraulic resistance is insignificant compared to the whole WTN, hydraulic resistance of the same can be ignored for the entire duration. Only the pressurized length (L-h) shown in Figure 2(b) is considered for modeling the immediate downstream pipe to the MBR. This assumption also eliminated the need to update the length of the immediate downstream pipe from the MBR after each time step.

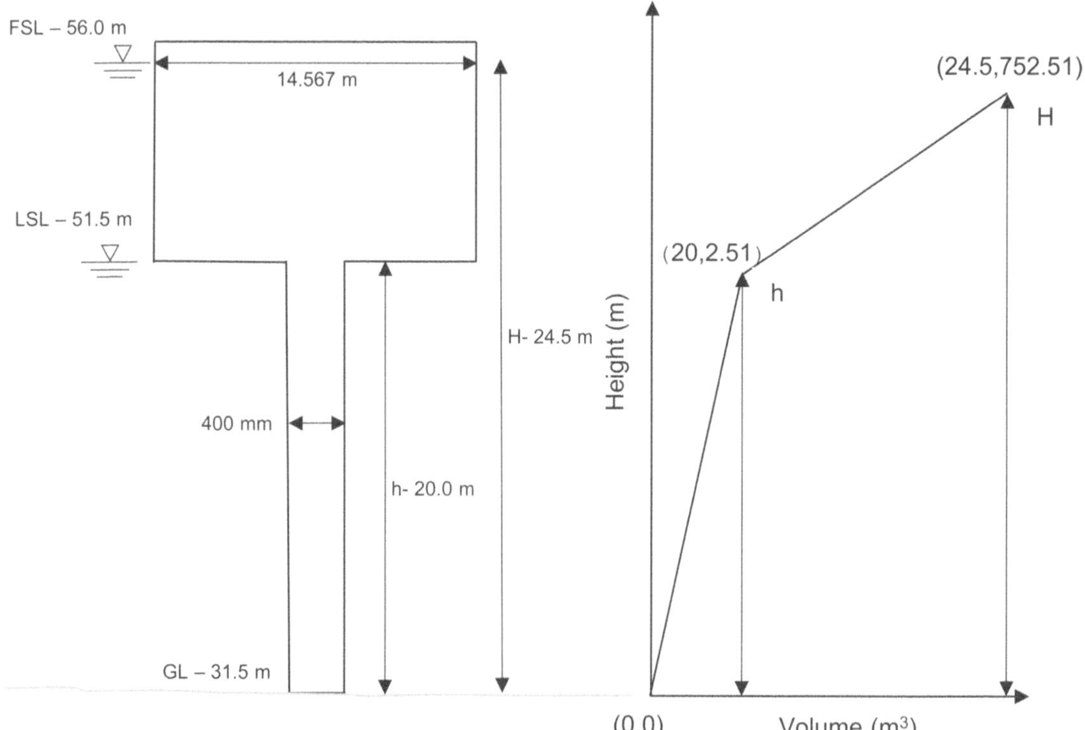


**Figure 3** | Shape of the tank with a volume curve for the proposed approach of modeling flow starvation (figure not to scale).

*Assumption III*: During operation, all changes are considered to occur under quasi-steady flow condition. Hence, smooth transition between steady-state simulation is able to depict the performance adequately.

*Assumption IV*: During entire supply period, valves to all SRs remain open as assumed in design stage, and all SRs try to draw water simultaneously irrespective of water received previously. This assumption allows to model extreme of poorly managed WTN.

The detailed methodology for implementing PDA-FSC is presented in Figure 4 and subsequent paragraphs describe the procedure.

The network upstream of the MBR (Sump–Pump–Pumping Main–MBR) is used to determine inflow to the MBR. Part of Figure 1 illustrates this network schematically. The physical pure water sump is represented as a reservoir element to replicate the constant water level in a pure water sump. The water discharges through the inlet of MBR at a constant level, as shown in Figure 1. The MBR is also represented as a reservoir element with the same elevation as the FSL of MBR. A pump curve derived from a single duty point (data obtained from pump manufacturer) is used to simulate the pump's performance. After obtaining a complete upstream network to the MBR (refer to Supplementary Material data SI-2/Badapokharan_MBR_Pumped network inflow.wtg), steady-state simulation is used to find the MBR inflow.

The water network downstream to the MBR (MBR-WTN-SR (Figure 1)) is used to determine the withdrawal capacity of the WTN. Typically, a reservoir element is used to simulate the water withdrawing tank of the WDN in a head-driven network (Batish 2003; Mohapatra *et al.* 2014; Sivakumar *et al.* 2020). The same approach is adopted, and a reservoir element and a control valve in the pipe upstream to the reservoir are used to replicate the water withdrawing behavior of the SR for simulation of the WTN as depicted in Figure 5(a) (Refer to Supplementary Material data SI-2/Badapokharan_MBR_withdrwal capacity curve.wtg). The control valve in the upstream pipe of SR prevents backward flow from the SR. A reservoir element represents the MBR to simulate the withdrawal capacity. Different withdrawal capacities (*W1*, *W2*, *W3*) at various water levels (i.e. FSL, LSL, GL of the MBR) are obtained by changing the elevation of the reservoir element representing the MBR. Since demand is not considered and flow depends on outlet properties, the proposed approach is based on the PDA-PDO approach used by Chandapillai (1991) where $H^{min}$ is considered equal to the elevation of the water discharge point ($H^{Avl} = H_j^{min} + K_j*(Q^{Avl})^n$).

After obtaining inflow and withdrawal capacities (*W1*, *W2*, *W3*), comparison is made to check the existence of FSC. If the inflow is greater than *W1* or *W2*, conventional PDA based on PDO should be enough to model the performance of WTN. If the inflow is between *W2* and *W3*, FSC exists, and SWC formation is possible between GL and LSL of the MBR. If the inflow is lower than *W3*, this method is not applicable, as SWC formation takes place below GL of the MBR.

After confirming the existence of FSC, the entire network (Sump–Pump–Pumping Main–MBR–WTN–SRs) is used for obtaining the performance of WTN which is depicted in Figure 5(b) (Refer to Supplementary Material data SI-2/EPS_Badapokharan_Operation as per design_base_model.wtg). Here, MBR and its immediate vertical downstream pipe are represented by the tank element (irregular-shaped tank), as depicted in Figure 3. Representation of SRs, in this case, remains the same as that of the model used for withdrawal. The MBR is a tank that fills from top, i.e. a tank inlet that discharges above the water surface. It can be modeled as a tank's inlet consisting of a pressure sustaining valve followed by a short length of large diameter pipe in EPANET. A detailed procedure for the same has been described by Rossman (2000). In this study, WaterGEMS is used and an equivalent EPANET model is included in Supplementary data SI-2 (Refer to Supplementary Material data SI-2/EPS_Badapokharan_Operation as per design_base_model.net). WaterGEMS can model the top-filling behavior as it has a built-in functionality in the tank element that provides a separate inlet at any tank level. For simulating the performance of the WTN in all stages of supply (i.e. normal, transition, equilibrium), extended period simulation (EPS) is done. To ensure that the simulation covers performance over all supply stages, the FSL is set as the starting boundary condition of operation, i.e. the starting water level in the MBR. The time step of the simulation is kept small so that the performance of the WTN during Stage 2 is captured. Multiple time steps are tried to check the convergence and stability of the results. The time step value for the initial trial is selected based on the withdrawal capacity of *W2* and the volume of the vertical portion of the immediate downstream pipe. The initial time step is selected to be lower than the time taken to empty the vertical section of the immediate downstream pipe, considering the constant withdrawal capacity *W2*. This process serves as a rule of thumb for selecting the time step. Once results converge, the stability of the result for Stage 1 and Stage 3 is established by observing the water level in the MBR and SWC's height with a lower time step.

After PDA-FSC simulation, the design demand for each SR is obtained based on the downstream population, per-capita demand, and the WTN operating hours (CPHEEO 1999). As all flows should match estimated or design demand for

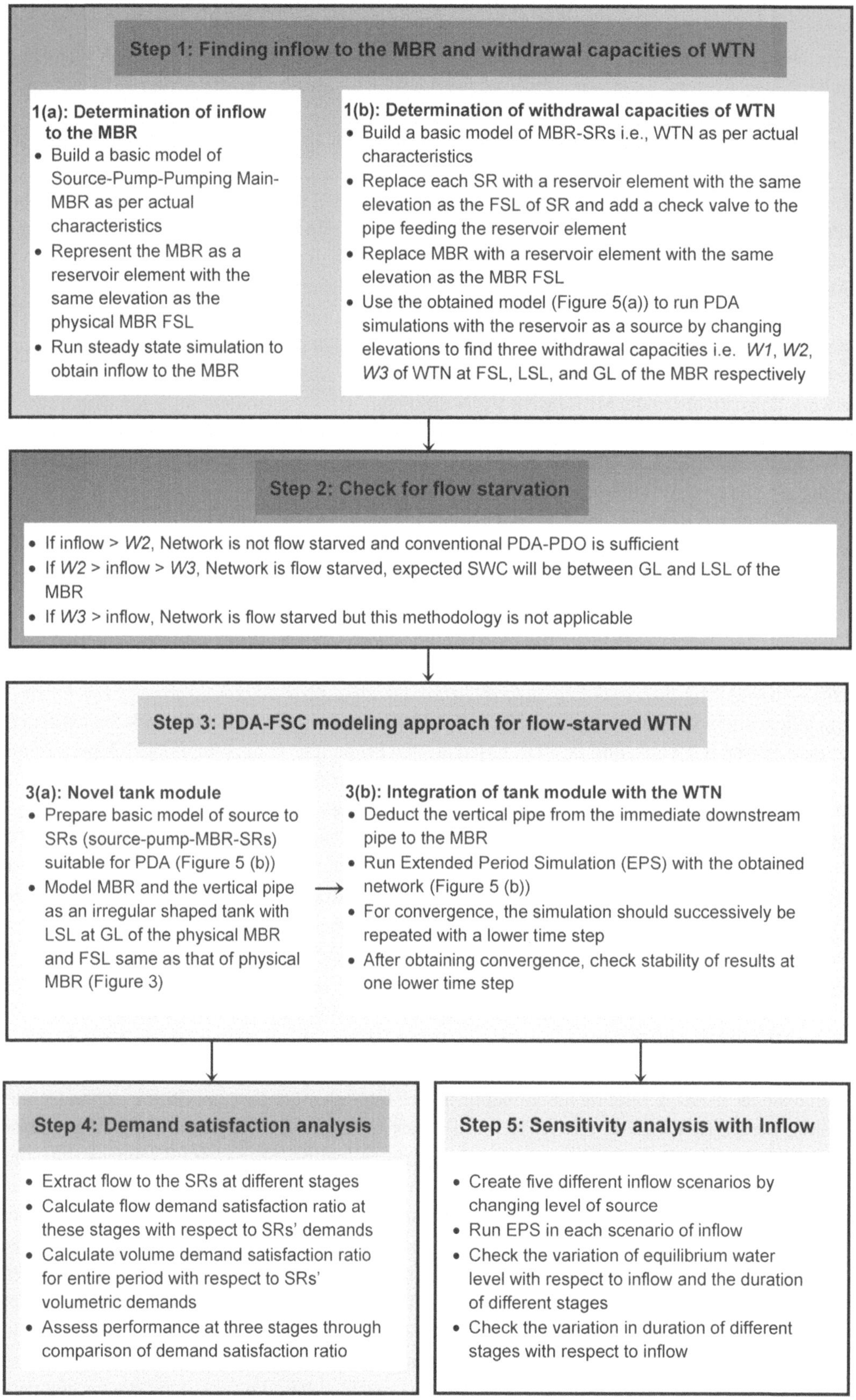

**Step 1: Finding inflow to the MBR and withdrawal capacities of WTN**

**1(a): Determination of inflow to the MBR**
- Build a basic model of Source-Pump-Pumping Main-MBR as per actual characteristics
- Represent the MBR as a reservoir element with the same elevation as the physical MBR FSL
- Run steady state simulation to obtain inflow to the MBR

**1(b): Determination of withdrawal capacities of WTN**
- Build a basic model of MBR-SRs i.e., WTN as per actual characteristics
- Replace each SR with a reservoir element with the same elevation as the FSL of SR and add a check valve to the pipe feeding the reservoir element
- Replace MBR with a reservoir element with the same elevation as the MBR FSL
- Use the obtained model (Figure 5(a)) to run PDA simulations with the reservoir as a source by changing elevations to find three withdrawal capacities i.e. $W1$, $W2$, $W3$ of WTN at FSL, LSL, and GL of the MBR respectively

**Step 2: Check for flow starvation**

- If inflow > $W2$, Network is not flow starved and conventional PDA-PDO is sufficient
- If $W2$ > inflow > $W3$, Network is flow starved, expected SWC will be between GL and LSL of the MBR
- If $W3$ > inflow, Network is flow starved but this methodology is not applicable

**Step 3: PDA-FSC modeling approach for flow-starved WTN**

**3(a): Novel tank module**
- Prepare basic model of source to SRs (source-pump-MBR-SRs) suitable for PDA (Figure 5 (b))
- Model MBR and the vertical pipe as an irregular shaped tank with LSL at GL of the physical MBR and FSL same as that of physical MBR (Figure 3)

**3(b): Integration of tank module with the WTN**
- Deduct the vertical pipe from the immediate downstream pipe to the MBR
- Run Extended Period Simulation (EPS) with the obtained network (Figure 5 (b))
- For convergence, the simulation should successively be repeated with a lower time step
- After obtaining convergence, check stability of results at one lower time step

**Step 4: Demand satisfaction analysis**

- Extract flow to the SRs at different stages
- Calculate flow demand satisfaction ratio at these stages with respect to SRs' demands
- Calculate volume demand satisfaction ratio for entire period with respect to SRs' volumetric demands
- Assess performance at three stages through comparison of demand satisfaction ratio

**Step 5: Sensitivity analysis with Inflow**

- Create five different inflow scenarios by changing level of source
- Run EPS in each scenario of inflow
- Check the variation of equilibrium water level with respect to inflow and the duration of different stages
- Check the variation in duration of different stages with respect to inflow

**Figure 4** | Flow chart for the methodology of the present study.

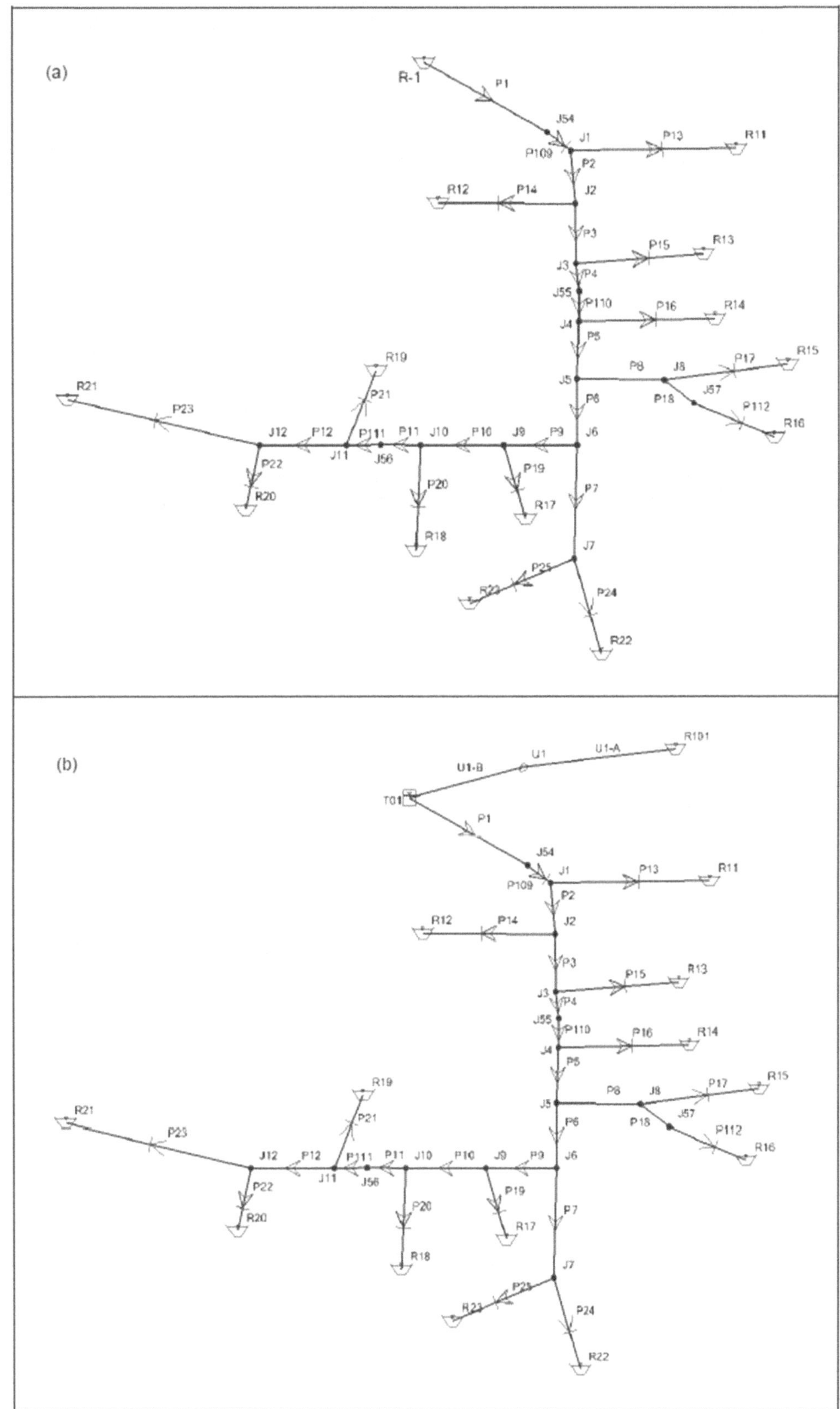



**Figure 5** | Network configuration used for (a) withdrawal capacity and (b) EPS of the flow-starved transmission network.

equity in practice, the Flow-based Demand Satisfaction Ratio (FDSR) at SRs is defined as the ratio of delivered flow (LPS) to required demand (LPS) at any instant. Similarly, volume-based DSR (VDSR) at SRs is defined as the ratio of delivered volume ($L$) to required volumetric demand ($L$) during a time period. FDSR values greater than one imply the SR receiving greater flow than designated demand. Similarly, a FDSR value of less than one is an indication that the SR is receiving less flow. The values of VDSR can be interpreted similarly for volumetric water demand. In this context, 'Equality' in WTN means each SR receives same flow or volume of water, whereas 'Equity' implies that each SR receive water proportional to their designated demand. FDSR and VDSR can serve as suitable indicators of equity for particular stage or period, respectively. The goal of operation should be minimizing the difference among SRs' satisfaction indicators, thereby improving equity. These indicators can be used to understand the real-life implication of WTN under FSC. Consequently, FDSR or VDSR is used for the performance evaluation of WTN in terms of equitable water distribution.

Since the performance of the flow-starved WTN is highly dependent on the MBR inflow, the impact of the MBR inflow variations on the performance of the WTN is examined. Through this examination, the robustness of the modeling approach is also tested. Drivers for the increase and decrease in inflow are investigated to cover the entire spectrum, and their estimates were found. An increase in inflow can be due to various reasons, such as larger pumps for future-proofing ($+20\%$) and flow increase due to increased water level in the sump ($+7\%$). The major reason for the decrease in inflow (up to $-20\%$) can be a leakage in the main or lower water availability at the source during summertime. Hence $\pm20\%$ is considered as a possible range of inflow variation, and five scenarios were created within a 20% deviation of base inflow simulation. The five scenarios are

1. Scenario 1: Inflow = base inflow
2. Scenario 2: Inflow = 0.8 times base inflow
3. Scenario 3: Inflow = 0.9 times base inflow
4. Scenario 4: Inflow = 1.1 times base inflow
5. Scenario 5: Inflow = 1.2 times base inflow

## Case study

A WTN from Badapokharan, Maharashtra, India, is selected to demonstrate the proposed PDA-FSC modeling approach. The overview and layout of the network are presented in Supplementary Material, Table S1-1 and Figure S1-2, respectively. The node and pipe details are provided in Supplementary Material, Table S1-3 and Table S1-4, respectively (Supplementary data SI-2). The system represents a WTN with MBR as source and SRs as demand points. Figure 5(a) shows the network model configuration used to determine the WTN withdrawal. Figure 5(b) shows the network's layout used for EPS performance evaluation during the three supply stages of the WTN operation.

## The rationale for selecting the operational regime of the WTN

The present study aims to demonstrate the performance of WTN under poor institutional capacity of a water utility with ineffective demand-side management and operational controls. For such a situation, it is assumed that if consumers receive excess water, they continue to withdraw due to uncertainty of supply, which is one extreme of most cases in India (Burt & Ray 2014; Ghorpade *et al.* 2021). This leads to the withdrawal of total water received at each SR by consumers. On the supply side of SR, all SRs are considered to withdraw water from the MBR simultaneously, as contemplated in the design. This operational regime is regarded as the suitable scenario to demonstrate the performance of WTN under three supply stages (normal, transitional, and equilibrium) as the downstream system valve setting remains the same and hence the effects of water level variations are easily captured. Although the present study includes only one operational regime, other more realistic operational regimes such as each SR getting closure after receiving total daily demand, can be modeled if the correct control mechanism is implemented.

## Validation method for the PDA-FSC modeling approach

Mohan & Abhijith (2020) have suggested using the mass balance approach to validate simulation results of flow and pressure-starved conditions. However, such an approach is unnecessary for the PDA-FSC as it is implicitly taken care by the tank element in the modeling. Deployment of irregular shape tank in the model eliminates the need for modeling partial flow present in the vertical portion of the immediate downstream pipe to MBR. Assumptions are made in the conceptual framework of PDA-FSC to improve the ease of modeling. Assumptions I and II are based on the insignificant hydraulic resistance of the vertical portion of the immediate downstream pipe to MBR. The applicability of the PDA-FSC approach is verified by comparing the instantaneous value of withdrawal capacity and the approximation used in modeling. As described in the explanation of Stage 2, only the pressurized section of the immediate downstream pipe (Length equal to $L$-$x$ in Figure 2) offers hydraulic resistance. However, conventional software cannot modify network length at each time step during simulation. Hence, the present study uses an approximation for pressurized section and considers constant ($L$-$h$) length to be pressurized throughout the simulation.

Three estimates are obtained from simulation for the withdrawal capacity of WTN. Withdrawal capacities of WTN at different piezometric heads, represented by HGLs (from GL to FSL of MBR at 0.5-m interval) are found in two extreme scenarios of pressurization (at the extreme value of $x$, i.e. $h$ and 0).

1. *Approximation I:* where $L$-$x = L$-$h$
2. *Approximation II:* where $L$-$x = L$

In addition, the third estimate of withdrawal capacity is made with the exact length of the pressurized section of the immediate downstream pipe to MBR to obtain an instantaneous value of withdrawal capacities. Instantaneous withdrawal capacities and these approximations at different HGL of MBR are shown in Figure 6.

Figure 6 shows that the estimates of withdrawal capacity (Approximation II) are lower or equal to the withdrawal capacity (instantaneous value), and withdrawal capacity (Approximation I) are consistently higher or equal to the withdrawal capacity (instantaneous value). The difference between the three values of withdrawal capacities is abysmally low at all HGLs. Maximum errors in the estimation of instantaneous withdrawal capacity are $+0.85$ and $-0.17\%$ of instantaneous withdrawal capacity in Approximation I and Approximation II, respectively. Additionally, the instantaneous withdrawal capacity always lies within two approximations. Significant uncertainty of flow and pressure in the water system exist due to uncertainty associated with correct network parameters and demand loading conditions (Bargiela & Hainsworth 1989). Hence, error arising from these approximations are acceptable. The hydraulic model could not be verified experimentally due to inadequate monitoring of operating conditions (flow and pressure) and unpredictable operation schedule at case study location. An additional validation of the PDA-FSC approach for two-reservoir network (represented in Supplementary Material, Figure S1-5) is included in Supplementary data SI-1.

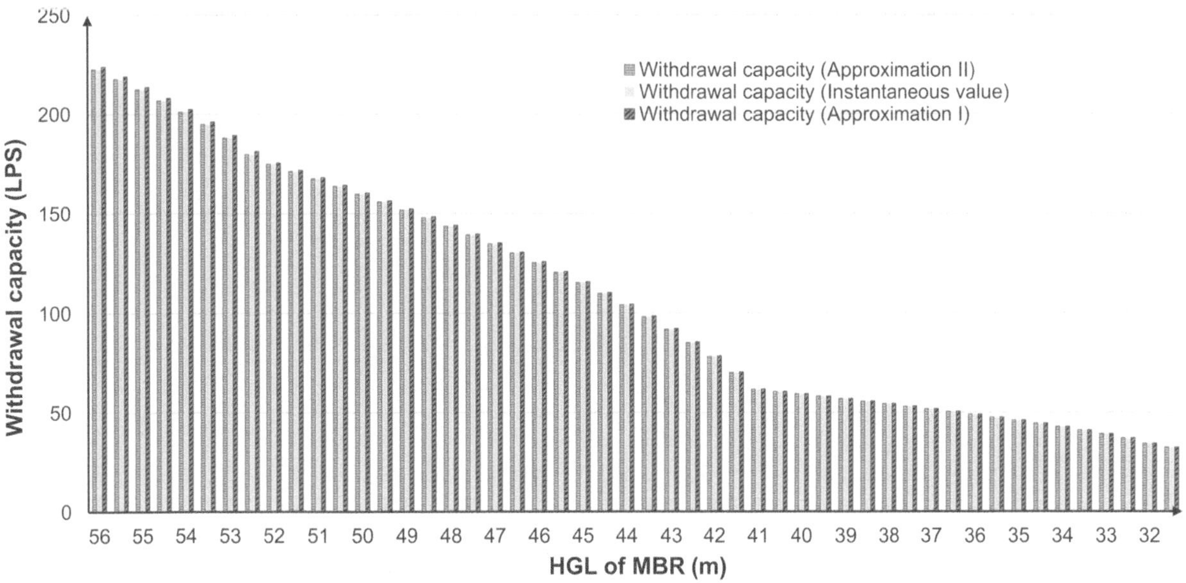


**Figure 6** | Comparison of the instantaneous value of withdrawal capacity and its approximations.

## RESULTS AND DISCUSSION

A steady-state simulation of the network upstream of the MBR is done to determine the inflow to the reservoir. The results show that the inflow is 47.44 LPS. The withdrawal capacities of the WTN at different piezometric heads governed by FSL (56 m), LSL (51.5 m), and GL (31.5 m) are found to be 224.84 LPS (*W1*), 171.20 LPS (*W2*), and 32.48 LPS (*W3*), respectively. From the comparison of inflow and withdrawal capacities, the WTN is assessed to be a flow-starved network because there is a formation of a SWC between GL and LSL of MBR. As described above, the calculation as per the rule of thumb shows that the starting time step should be less than 14 s. The EPS is done for the entire network (Figure 5(b) using a time step of 5 s for a duration of 86,400 s (24 h) to cover the mandated 16-h operation period of the WTN. As Stage 3 is attained within the period (at 5,700 s), the resulting net outflows and water levels in the MBR of Badapokharan WTN are plotted as time series as shown in Figure 7(a) and 7(b), respectively. The SWC is established at 3.96 m from the GL of the MBR. The EPS is repeated for a smaller time step of 1 s; no change is found in the SWC's height and water level in Stage 1, which confirms that the results during Stages 1 and 3 are independent of the time step. It is also observed that the water level and pressure during transition is dependent on time step and can be considered as a limitation of the PDA-FSC modeling approach. However, stable results are observed during most of the operating period and a stable water column obtained from different time steps shows that the hydraulic model can predict stable performance except for transition stage. Figure 7(b) shows that Stage 1 lasts 1.39 h and Stage 2 lasts only 0.08 h (less than 5 min). After attaining Stage 3, the water level in the MBR remains stable for the rest of the operational hours (Figure 7(b)). Since the WTN is usually operated for 16 h, it functions at Stage 3 for more than 14 h.

To determine the FDSR, the demand of SRs and flow toward SR are compared. Table 1 presents flows obtained from simulation under various supply stages and design demands. Observations for Stage 2 have not been included because it lasts only

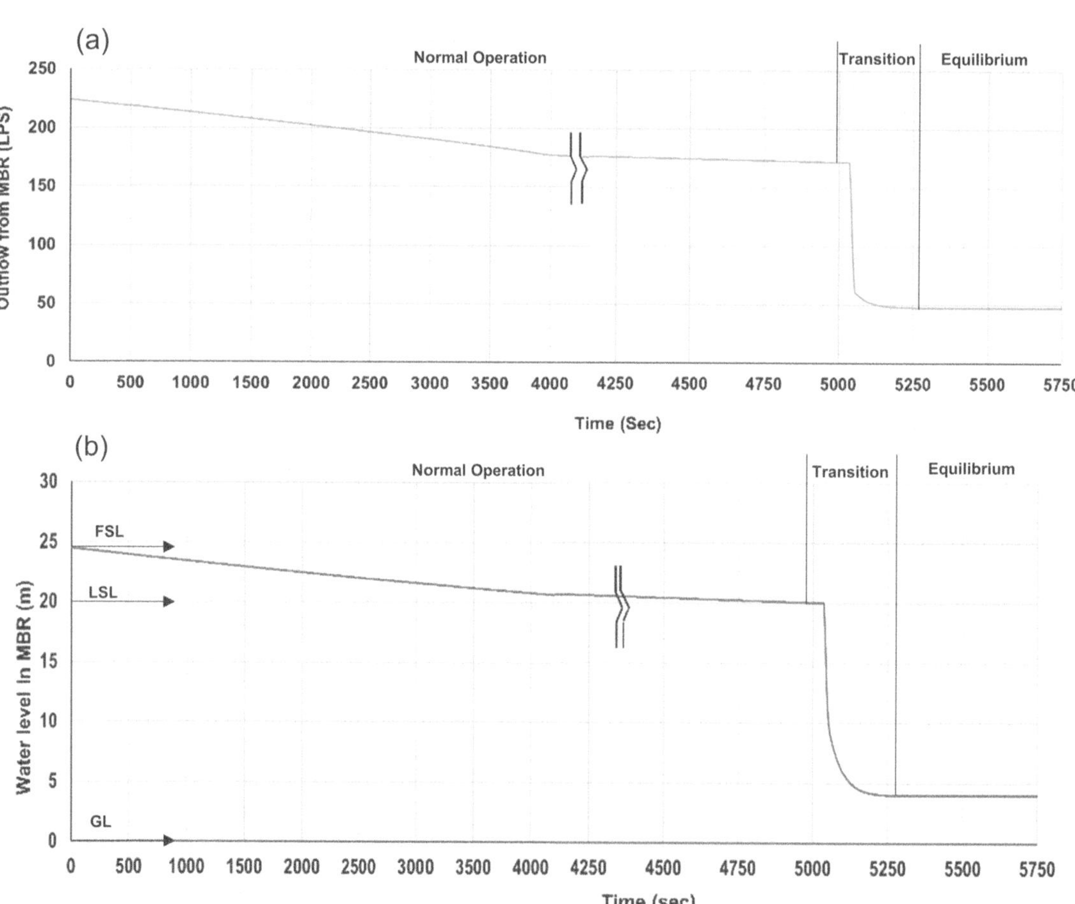


**Figure 7** | (a) The water level in the MBR at different supply stages and (b) outflow at the MBR at different stages of supply. Stage 1: normal operation, Stage 2: transition, Stage 3: equilibrium.

**Table 1** | Demand satisfaction under various supply stages

| S. No. | Village SR | Demand (LPS) | Flow at Stage 1 start (LPS) | Flow at Stage 1 end (LPS) | Flow at Stage 3 (LPS) | FDSR | VDSR |
|---|---|---|---|---|---|---|---|
| 1 | R13 | 2.43 | 13.75 | 13.53 | 7.60 | 3.13–5.64 | 3.34 |
| 2 | R12 | 2.43 | 127.62 | 108.48 | 0 | 0–52.52 | 4.26 |
| 3 | R11 | 6.25 | 37.57 | 0 | 0 | 0–6.01 | 0.20 |
| 4 | R23 | 2.43 | 9.36 | 9.25 | 6.45 | 2.65–3.85 | 2.76 |
| 5 | R22 | 2.43 | 0 | 0 | 0 | 0–0 | 0 |
| 6 | R21 | 2.43 | 0 | 0 | 0 | 0–0 | 0 |
| 7 | R20 | 2.43 | 4.25 | 4.23 | 3.69 | 1.52–1.75 | 1.54 |
| 8 | R19 | 5.21 | 0 | 0 | 0 | 0–0 | 0 |
| 9 | R18 | 1.39 | 3.84 | 3.83 | 3.57 | 2.57–2.76 | 2.59 |
| 10 | R17 | 1.39 | 19.05 | 18.93 | 16.18 | 11.64–13.71 | 11.83 |
| 11 | R16 | 3.82 | 0 | 0 | 0 | 0–0 | 0 |
| 12 | R15 | 5.21 | 0 | 0 | 0 | 0–0 | 0 |
| 13 | R14 | 2.43 | 13.83 | 13.68 | 9.95 | 4.09–5.69 | 4.23 |

for a few minutes. Observations based on FDSR during different supply stages reveal that the FDSR varied widely across the SRs and in various supply stages. As shown in Table 1, several SRs (R13, R23, R20, R18, R17, R14), typically with lower FSL (Supplementary Material, Figure S1-2), receive more flow than their design demand, and few SRs do not receive water (R22, R21, R19, R16, R15). These observations imply that flow distribution between SRs is not equitable. FDSR worsens after reaching Stage 3 as the number of SRs having zero inflow increases from 5 to 7, with a decrease in demand satisfaction for most of the other SRs. VDSR results reveal cumulative effect of entire supply period where several SRs (R13, R23, R20, R18, R17, R14) are overserved whereas the remaining SRs are underserved. The flow control devices are recommended to achieve equitable water distribution at pipe branch leading to overserved SRs. The goal of the WTN is to match supply and demand at the whole network and each individual SR. It is evident from Table 1 that the total WTN withdrawal under PDA is many folds higher than the design demand. Even under normal operation which have higher withdrawal, this increase in withdrawal is not distributed equally among SRs. This inequity demonstrates performance failure and need for the supply-side control to SRs.

## The sensitivity of performance with inflow variation

As inflow scenario creation can help assess the effect of inflow variations on the performance of the WTN, five inflow scenarios are created. For performance evaluation, the MBR's water level variations are tracked in all five scenarios and presented in Figure 8(a). For better visualization of the transition from Stage 1 to Stage 3, the water levels during the transition are shown in Figure 8(b). For any inflow, the water level in the MBR drops from Stage 1 and eventually attains equilibrium. A comparative assessment of water levels at Stage 3 for different flows shows that the SWC's height varies in the same direction as inflow. As expected, higher inflow can delay the formation of the SWC, and lower inflow can lead to the early formation of the SWC. The ability of the PDA-FSC modeling approach to simulate performance over a realistic range of inflow exhibits its suitability.

The PDA-FSC approach provides an opportunity to assess the effect of inflow variation on the WTN performance. Since insufficient inflow is a causal factor in flow starvation and leads to reduced flow at the downstream SRs, the reduced inflow can further enhance the formation of SWCs in SRs. The SWC formations in SRs often leads to incomplete demand satisfaction at the consumer end.

One of the significant problems identified through the PDA-FSC approach is FSC in the WTN. Three options exist for eliminating FSC (1) increasing MBR capacity, (2) increasing MBR inflow to match withdrawal and (3) reducing SRs' withdrawal. It is evident from the PDA-FSC results (Figure 7) that increasing the MBR capacity can elongate the duration of normal operation. Since normal operation lasts for only 8.64% (1.38 h out of 16 h) of the total duration, a multi-fold increase in capacity is needed to eliminate the SWC, which may not be the best economic option. From sensitivity analysis, increasing inflow by 20% results in a delay in the formation of the SWC; theoretically, increasing inflow can eliminate the appearance of the

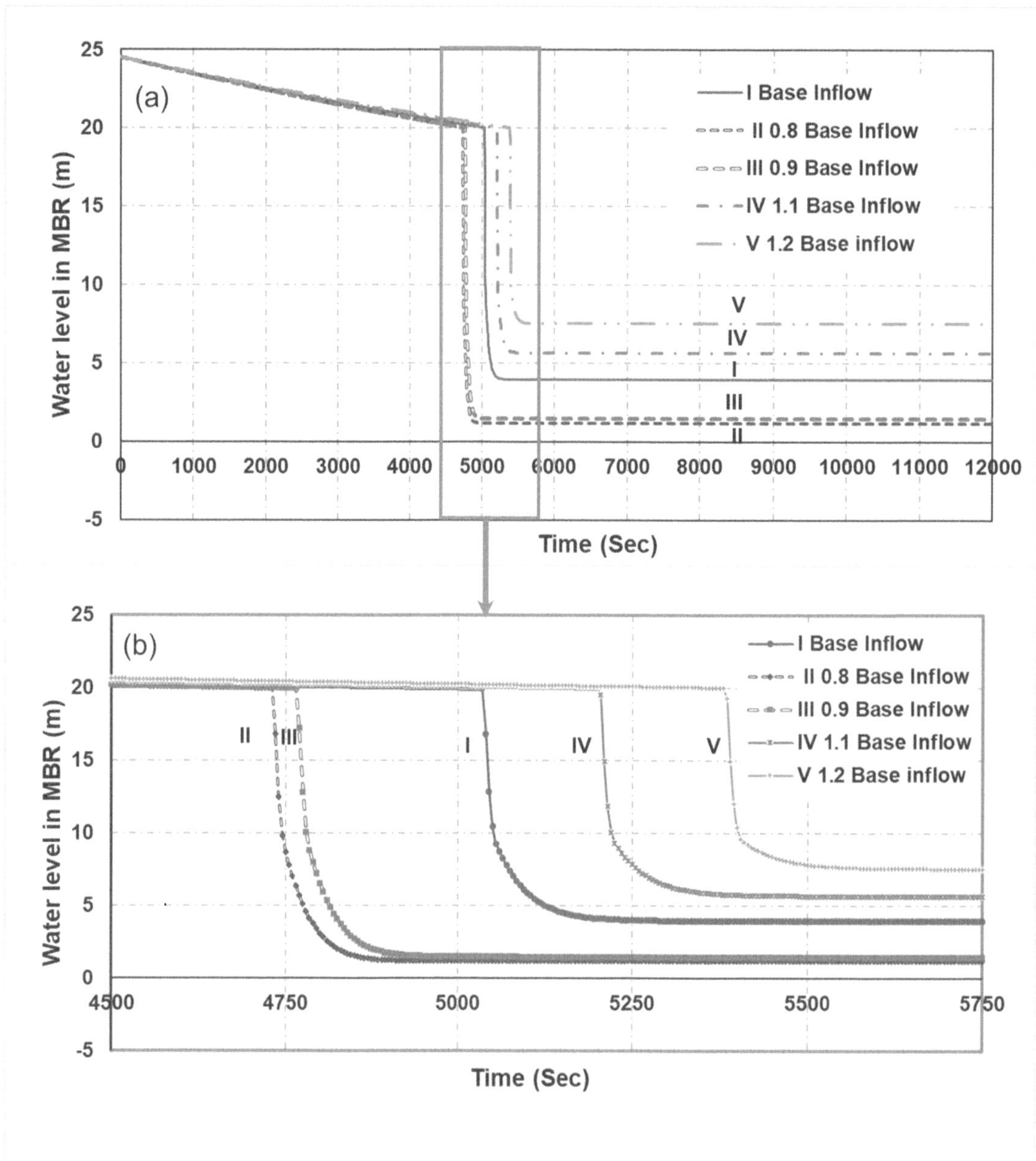

**Figure 8** | Water level in the MBR (a) in three supply stages and (b) during transition from Stage 1 to Stage 3.

SWC; however, it may not be economically feasible. Out of the three management options, reducing excessive SR withdrawal seems to be most economically viable option as it involves minor infrastructural changes such as control valves and equipment as opposed to major rehabilitations of infrastructure needed for other options. The option of controlling withdrawal also addresses the problem on the downstream side of MBR and creates a better opportunity for equitable distribution.

## Managerial implications of the PDA-FSC approach

The developed PDA-FSC approach helps to understand the implications of poor inlet pressure and demand management during operation. The results of the PDA-FSC approach also help in attributing the cross-dependence of different WSS components (pumping main, MBR, WTN). Generally, the additional capacity of the WTN is considered an advantage for WSS operation. However, the present study demonstrates the downside of unnecessary additional capacity of the WTN and how it can create inequitable water distribution in poorly controlled WTN.

The PDA-FSC approach can incorporate several prevailing operational practices of the IWS. The modeling approach provide an opportunity to reflect on the control system of the WTN and create an avenue for designing the operation schedule according to the topography and demand. Several control equipments (flow control valve, pressure reducing valve, orifice, and manifold) have been advocated for effective pressure management, i.e. to match supply and demand at both network and individual SR levels. This modeling approach can assess the control equipment's effectiveness in achieving the desired goal. The PDA-FSC approach can be used as a management tool to verify or devise operation schedules for different inflow scenarios based on prevailing operation circumstances such as operator-based operation and valve throttling.

The study demonstrates the importance of operational control based on hydraulics in this system. It may thus lead to a policy focus on operation instead of having bigger pipes through public investment in gravity-fed WTN. The present study shows that achieving equitable supply is related to planning as well as the operation of the network. The PDA-FSC modeling approach used in the study is easy to replicate with a low learning curve. This is vital for developing countries where comprehensive modeling is seldom used for designing operation schedules.

## CONCLUSIONS

The present study highlights the limitations of the existing hydraulic modeling approaches for flow-starved conditions in water networks, a common phenomenon in the IWS system prevailing in most developing countries. The modeling approach developed in this study provides a better and easier way to simulate flow-starved conditions in the WTN. The methodology developed has been implemented on a WTN in India. The results from this modeling approach show that redundant withdrawal capacity in an inadequately controlled WTN can lead to inequitable water distribution indicating performance failure. The study further demonstrates the impact of a flow-starved network and its inability to fulfill the downstream service reservoir's demands. The results from the developed modeling approach can be used to derive better operational schedule for IWS.

## ACKNOWLEDGMENTS

Authors acknowledge the funding received under the sponsored research project from the Department of Science and Technology (DST), India – Water Technology Initiative (WTI) – 2017 DST/TM/WTI/2K17/39. The authors acknowledge all the cooperation and support provided by Palghar Zilla Parishad, Palghar, Maharashtra, India. The authors also acknowledge editor and two anonymous reviewers for insightful and comprehensive suggestions.

## DATA AVAILABILITY STATEMENT

All relevant data are available from an online repository or repositories (https://dx.doi.org/10.5281/zenodo.7162154).

## CONFLICT OF INTEREST

The authors declare there is no conflict.

## REFERENCES

Agrawal, M. L., Gupta, R. & Bhave, P. R. 2007 Optimal design of level 1 redundant water distribution networks considering nodal storage. *Journal of Environmental Engineering* **133** (3), 319–330. doi:10.1061/(asce)0733-9372(2007)133:3(319).

AWWA 2008 *Distribution System Requirements for Fire Protection. 4th edn, American Water Works Association*, 4th edn. American Water Works Association, Denver, USA.

Bargiela, A. & Hainsworth, G. D. 1989 Pressure and flow uncertainty in water systems. *Journal of Water Resources Planning and Management* **115** (2), 212–229. doi:10.1061/(ASCE)0733-9496(1989)115:2(212).

Batish, R. (2003) A new approach to the design of intermittent water supply networks. In: *World Water & Environmental Resources Congress 2003*. Reston, VA: American Society of Civil Engineers, pp. 1–11. doi:10.1061/40685(2003)123.

Bentley 2022 *Elements and Element Attributes, Bentley WaterGEMS CONNECT Edition Help*. Available from: https://docs.bentley.com/LiveContent/web/Bentley WaterGEMS SS6-v1/en/GUID-D1C8970FA3FF4D0291447CE2B68A01C7.html (accessed 17 February 2022).

Bhave, P. R. 1981 Node flow analysis distribution systems. *Transportation Engineering Journal of ASCE* **107** (4), 457–467. doi:10.1061/TPEJAN.0000938.

Bhave, P. R. & Gupta, R. 2000 Design, performance and operation of regional rural water supply systems. *Journal/Indian Water Works Association* **4**, 273–278.

Burt, Z. & Ray, I. 2014 Storage and non-payment: persistent informalities within the formal water supply of hubli-dharwad, India. *Water Alternatives* **7** (1), 106–120.

Cembrowicz, R. G. & Ates, S. 1998 The water supply network analysis tool KANET. *WIT Transactions on Ecology and the Environment* **19**. Available from: https://www.witpress.com/Secure/elibrary/papers/HY98/HY98009FU.pdf.

Chandapillai, J. 1991 Realistic simulation of water distribution system. *Journal of Transportation Engineering* **117** (2), 258–263. doi:10.1061/(ASCE)0733-947X(1991)117:2(258).

Charalambous, B. & Laspidou, C. 2017 *Dealing with the Complex Interrelation of Intermittent Supply and Water Losses, IWA Publishing*. IWA Publishing, London, UK. doi:10.2166/9781780407074.

Conety Ravi, S., Thurvas Renganathan, N., Perumal, S. & Paez, D. 2019 Analysis of water distribution network under pressure-deficient conditions through emitter setting. *Drinking Water Engineering and Science* **12**, 1–13. https://doi.org/10.5194/dwes-12-1-2019.

CPHEEO 1999 *Manual on Water Supply and Treatment. 3rd edn, Central Public Health and Environmental Engineering Organization, Minstry of Urban Development, Government of India*, 3rd edn. Minstry of Urban Development, Government of India, New Delhi, India. Available from: http://cpheeo.gov.in/cms/manual-on-water-supply-and-treatment.php.

Farmani, R., Dalton, J., Charalambous, B., Lawson, E., Bunney, S. & Cotterill, S. 2021 Intermittent water supply systems and their resilience to COVID-19: IWA IWS SG survey. *Journal of Water Supply: Research and Technology-Aqua* **70** (4), 507–520.

Fujiwara, O. & Li, J. 1998 Reliability analysis of water distribution networks in consideration of equity, redistribution, and pressure-dependent demand. *Water Resources Research* **34** (7), 1843–1850.

Ghorpade, A., Sinha, A. K. & Kalbar, P. P. 2021 Drivers for intermittent water supply in India: critical review and perspectives. *Frontiers in Water* **3** (September), 1–15. doi:10.3389/frwa.2021.696630.

Gupta, R. & Bhave, P. R. 1996 Comparison of methods for predicting deficient-network performance. *Journal of Water Resources Planning and Management* **122** (3), 214–217. doi:10.1061/(ASCE)0733-9496(1996)122:3(214).

Ingeduld, P., Pradhan, A., Svitak, Z. & Terrai, A. 2008 Modelling intermittent water supply systems with EPANET. In: *Water Distribution Systems Analysis Symposium 2006*. pp. 1–8. https://doi.org/10.1061/40941(247)37.

Kalbar, P. & Gokhale, P. 2019 Decentralized infrastructure approach for successful water supply systems in India: use of multi-outlet tanks, shafts and manifolds. *Journal of Water Supply: Research and Technology-Aqua* **68** (4), 295–301. doi:10.2166/aqua.2019.158.

Kalbar, P. P., Kulkarni, V. & Gokhale, P. 2014 Role of hydraulic modeling in development of road map for achieving $24 \times 7$ continuous water supply. In *46th IWWA Convention - 2014, Bangalore*, pp. 175–182.

Mahmoud, H. A., Savić, D. & Kapelan, Z. 2017 New pressure-driven approach for modeling water distribution networks. *Journal of Water Resources Planning and Management* **143** (8), 04017031. doi:10.1061/(ASCE)WR.1943-5452.0000781.

Meyer, D. & Ahadzadeh, N. 2021 Discussion of 'Hydraulic analysis of intermittent water-distribution networks considering partial-flow regimes' by S. Mohan and G. R. Abhijith. *Journal of Water Resources Planning and Management* **147** (11), 2–4. doi:10.1061/(ASCE)WR.1943-5452.0001466.

Mohan, S. & Abhijith, G. R. 2020 Hydraulic analysis of intermittent water-distribution networks considering partial-Flow regimes. *Journal of Water Resources Planning and Management* **146** (8), 04020071. doi:10.1061/(ASCE)WR.1943-5452.0001246.

Mohapatra, S., Sargaonkar, A. & Labhasetwar, P. K. 2014 Distribution network assessment using EPANET for intermittent and continuous water supply. *Water Resources Management* **28** (11), 3745–3759. doi:10.1007/s11269-014-0707-y.

Randeniya, A., Radhakrishnan, M., Sirisena, T. A. J. G., Masih, I. & Pathirana, A. 2022 Equity-performance trade-off in water rationing regimes with domestic storage. *Water Supply* **22** (5), 4781–4797. https://doi.org/10.2166/ws.2022.188.

Reddy, L. S. & Elango, K. 1989 Analysis of water distribution networks with head-dependent outlets. *Civil Engineering Systems* **6** (3), 102–110. doi:10.1080/02630258908970550.

Reddy, L. S. & Elango, K. 1991 A new approach to the analysis of water straved network. *Journal of Indian Water Works Association* **1**, 31–38.

Rossman, L. A. 2000 *EPANET 2.0 User's Manual, United States Environmental Protection Agency*. Cincinnati, OH, USA. Available from: https://nepis.epa.gov/Adobe/PDF/P1007WWU.pdf.

Sarisen, D., Koukoravas, V., Farmani, R., Kapelan, Z. & Memon, F. A. 2022 Review of hydraulic modelling approaches for intermittent water supply systems. *Journal of Water Supply: Research and Technology-Aqua* **71**, 1291–1310. https://doi.org/10.2166/aqua.2022.028.

Sashikumar, N., Mohankumar, M. S. & Sridharan, K. 2003 Modelling an intermittent water supply. *World Water and Environmental Resources Congress* 2043–2052. doi:10.1061/40685(2003)261.

Shirzad, A. 2020 A model for pressure driven analysis-design of water distribution networks. *Journal of Applied Water Engineering and Research* **8** (2), 79–87. doi:10.1080/23249676.2020.1761895.

Simukonda, K., Farmani, R. & Butler, D. 2018 Intermittent water supply systems: causal factors, problems and solution options. *Urban Water Journal* **15** (5), 488–500. doi:10.1080/1573062X.2018.1483522.

Sivakumar, P., Gorev, N. B., Tanyimboh, T. T., Kodzhespirova, I. F., Suribabu, C. R. & Neelakantan, T. R. 2020 Dynamic pressure-dependent simulation of water distribution networks considering volume-driven demands based on non-iterative application of EPANET 2. *J Water Resour Plan Manag* **146** (6), 06020005. https://doi.org/10.1061/(ASCE)WR.1943-5452.0001220.

Tabesh, M., Tanyimboh, T. & Burrows, R. 2001 Head-driven simulation of water supply networks. *International Journal of Engineering – Transactions A: Basics* **15** (1), 11–22.

Taylor, D. D. J., Slocum, A. H. & Whittle, A. J. 2019 Demand satisfaction as a framework for understanding intermittent water supply systems. *Water Resources Research* **55** (7), 5217–5237. doi:10.1029/2018WR024124.

Wagner, B. J. M., Shamir, U. & Marks, H. 1988 Water distribution reliability: simulation methods I. *Journal of Water Resources Planning and Management* **114** (3), 253–275.

Walski, T. M., Chase, D. V., Savic, D. A., Grayman, W. M., Beckwith, S. & Koelle, E. 2003 *Advanced Water Distribution Modeling and Management*. Bentley Institute Press, Waterbury, CT.

First received 21 February 2022; accepted in revised form 8 December 2022. Available online 20 December 2022

IWA Publishing's authorised EU representative for General Product
Safety Regulations is Diane D'Arras, 15 rue Duret, 75116 Paris,
France, e-mail: safety@iwap.co.uk.

Printed and bound by CPI Group (UK) Ltd, Croydon, CR0 4YY
12/05/2026
02108874-0002